Lecture Notes in Economics and Mathematical Systems

499

Springer
Berlin
Heidelberg
New York
Barcelona
Hong Kong
London
Milan
Paris
Singapore
Tokyo

André Casajus

Focal Points
in Framed Games

Breaking the Symmetry

Springer

Author

André Casajus
Universität Hohenheim
Institut für Agrarpolitik und landwirtschaftliche Marktlehre
70593 Stuttgart, Germany

Cataloging-in-Publication data applied for

Die Deutsche Bibliothek - CIP-Einheitsaufnahme

Casajus, André:
Focal points in framed games : breaking the symmetry / André Casajus.
- Berlin ; Heidelberg ; New York ; Barcelona ; Hong Kong ; London ;
Milan ; Paris ; Singapore ; Tokyo : Springer, 2001
  (Lecture notes in economics and mathematical systems ; 499)
ISBN 978-3-540-41406-3   ISBN 978-3-642-51112-7 (eBook)
DOI 10.1007/978-3-642-51112-7

ISSN 0075-8450

Springer-Verlag Berlin Heidelberg New York
a member of BertelsmannSpringer Science+Business Media GmbH

© Springer-Verlag Berlin Heidelberg 2001
Softcover reprint of the hardcover 1st edition 2001

Typesetting: Camera ready by author
Printed on acid-free paper   SPIN: 10791409   55/3142/du   5 4 3 2 1 0

Für meine Eltern

Acknowledgments

This book was accepted as doctoral dissertation by the Faculty of Economics (*Wirtschaftswissenschaftliche Fakultät*) of the University of Leipzig. During my time in Leipzig, I received various kinds of support in developing the ideas set forth in this book. First of all, I would like to thank Harald Wiese, my principal advisor, for his permanent encouragement and a lot of helpful discussions on the subject. I also benefitted from discussions with Arnis Vilks (Leipzig Graduate School of Management) and Thomas Voss (Institute of Sociology) and with my colleagues Dirk Bültel, Michael Geisler, Matthias Maul, and Markus Wimmer.

Early papers underlying this book were presented at the 17th Arne Ryde Symposium (August 21-22, 1997, Lund, Sweden) and the 3rd Spanish Meeting on Game Theory and Applications (June 15-17, 1998, Barcelona, Spain). I extend my thanks to the participants of both conferences for their critical comments. Chapter 2 is based on my paper "Focal Points in Framed Strategic Forms" (Games and Economic Behavior 32, 2000: 263–291). I thank an anonymous referee for valuable suggestions which considerably improved both the paper and the chapter.

The last steps of finishing my dissertation were made at the University of Hohenheim where I also prepared the manuscript. Many thanks to Tilman Becker who generously permitted me to spent enough time on this project. Thanks also to my colleagues in Hohenheim who helped me to settle down that far from my beloved Leipzig.

Last but not least, I am indebted to my former wife Isabel who patiently attended this endeavor for quite a long time.

Hohenheim, September 2000   *André Casajus*

Contents

1 Introduction

Go to, let us go down, and there confound their language, that they may not understand one another's speech.
So the LORD scattered them abroad from thence upon the face of all the earth: and they left off to built the city. (Genesis 11.7–8)

1.1 Static Focal Points

1.1.1 Coordination

In real life, people quite often face situations in which they prefer to act in the *same* way, but they are not particular about the *concrete* way of acting. Some examples are given below:

Credit cards: Buyers want to possess the credit cards potential sellers do accept. Also, sellers wish to have contracted the credit card company the credit cards of which potential customers usually carry along. For both, basically, it is all the same which credit card this is. What matters is that both choices coincide.

Communication, information transmission: The transmission of information requires that the signals used have the same meaning to both the sender and the receiver. But it is inessential which signal has a certain meaning. In verbal communication, this basically means that the people involved use the same (natural) language, though even then some ambiguities remain. Things are a bit more difficult for non-verbal communication, for example data transmission between computers—both sides have to use the same or compatible protocols. Again, what matters is the protocols to be the same for both sides.

Road traffic: Encounting cars on a road clearly prefer to drive on the same side of the road. As the different rules in Great Britain and Germany show, driving on the right-hand side is as good as driving on the left-hand side—as long as all drivers do this.

2

	s_{21}	s_{22}	s_{23}
s_{11}	1 / 1	0 / 0	0 / 0
1 s_{12}	0 / 0	1 / 1	0 / 0
s_{13}	0 / 0	0 / 0	1 / 1

Fig. 1.1. A pure coordination game

Meeting place problems: Two persons want to meet each other, and all what matters is that they succeed in doing so—they are not particular about the meeting place itself.

The underlying structure of all these situations is that of a pure coordination problem. In real life, people quite often are successful in solving these kinds of coordination problems. To explain this success, however, is a major puzzle for game theory.

In game theory, problems of pure coordination are represented by strategic games as in Figure 1.1. There are two players—1 and 2—who wish to coordinate on one of three locations. This is modelled by the three strategies for each of the players, where the strategy s_{ik} stands for player i's choosing the kth location. The players' preference for coordination—choosing the same location—is represented by payoffs that are associated with the combinations of the players' strategies. In case of coordination, the players get the payoff 1, and in case of discoordination, they get the payoff 0.

In a sense—to be made precise in Section 2.3, the players and their strategies are somehow symmetric in this game. Besides the labels, for example 1 or s_{11}, players and strategies cannot be distinguished by the game theorist. Since there is much arbitrariness in choosing these labels, one could argue that they *should* not be distinguished by the game theorist: Whenever a game theorist finds an argument to suggest a particular strategy to a player, the same kind of argument also applies to this player's other strategies. Therefore, the game theorist's recommendation for playing the game is randomization—both players' choosing all of their strategies with the same probability. Following this recommendation, the players coordinate with a probability of $\frac{1}{3}$. Without breaking the symmetry of the strategies, better outcomes cannot be supported by game theory.

In real life as well as in experiments, however, people do much better. In contrast to the game theorist, the way in which the players perceive the strategies and describe them to themselves is far from being arbitrary. So the real-world players can exploit these descriptions in order to coordinate. In order to capture this ability to coordinate, Schelling (1960) introduces the concept of focal points. He argues that players quite often can find some

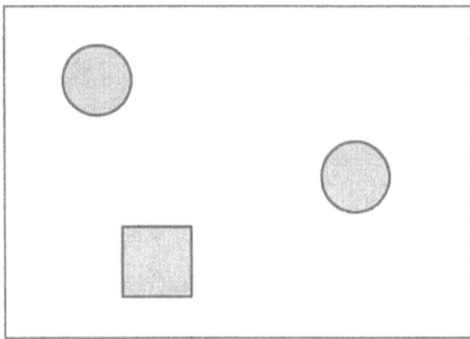

Fig. 1.2. Pure coordination game #1

prominent and unique clue—a focal point—in their perception of the game that enables them to coordinate. This is explained with an example.

Consider a pure coordination game in which two players independently have to choose one of three physical objects. They get a prize if they take the same object, otherwise they get nothing. Let the objects be those shown in Figure 1.2—two balls and one cube. Intuitively, it is clear that the players should take the cube—the cube stands out and enables the players to coordinate for sure, while considering to take a ball involves the question which one to choose. Hence, the cube is the focal point of this game: for players with good vision, the objects' shapes are prominent, and the cube is unique. Since the strategic game that represents this coordination problem is the same as given in Figure 1.1, however, this focal point cannot be explained within the framework of traditional game theory.

Consider now the alternative object set in the Figure 1.3. Here, the clue for coordinating is the color (brightness) of the balls, and the obvious focal point is the dark ball. As the previous game, this game can represented by the strategic game in Figure 1.1. Nevertheless, supposing that the game theorist numbers the strategies according to the location of corresponding objects, the focal point arguments select different strategies.

The examples above reveal that traditional game theory cannot explain focal points in pure coordination games, i.e., a class of games that are characterized by a high degree of symmetry between their strategies, because the players' description of the strategies is not formalized within these games. Therefore, it is desirable to incorporate these descriptions into game theory in a general way. i.e.,

- to extend strategic games by a structure that represents the players' description of a game, and
- to introduce a solution concept that
  - exploits this structure in order to break some of the symmetries in strategic games

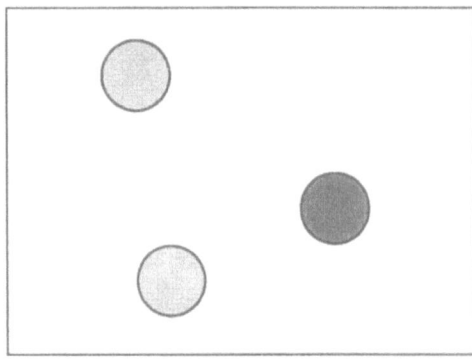

Fig. 1.3. Pure coordination game #2

○ in accordance with our intuitions about focal points.

This is what Chapter 2 is trying to accomplish: We introduce framed strategic games that extend strategic games by a structure called frame which formalizes the players' perception of their strategies in a general way. Making up these frames, some degree of arbitrariness in choosing the denotations used to represent the players' descriptions remains. For example, an English speaking game theorist probably will use a different term to denote a players' perception the redness of an object to choose—the label **red**—than an German speaking game theorist—the label **rot**. So, we must be very careful in interpreting and exploiting these labels. This is formalized with Definition 2.5.5 (p. 42).

Of course, this aim is a limited one. As argued in Subsection 1.1.2, the restriction to the symmetry-breaking feature of focal points prevents the explanation of focal points in bargaining. In addition, pure coordination games are not the only, but the main field of application for symmetry-breaking focal points (see next subsection). Sugden (1995, p. 534) justifies this restriction as follows: "It might be objected that pure coordination games are highly artificial, and that in games of real life we almost always find some degree of payoff asymmetry, repetition, or communication. But this, although true, is not an adequate reason for ignoring pure coordination games. These games may be thought of as controlled experiments which, by filtering out other features of real-word games that might induce players to choose one strategy rather than another, isolate the effects of labeling."

1.1.2 Bargaining

The pure coordination games considered above were characterized by identical preferences of the players. In contrast, in bargaining situations, besides the common interest to reach an agreement, the players have conflicting interests concerning the agreement itself. This is explained by help of an example:

	s_{21}	s_{22}	s_{23}	s_{24}
s_{11}	0 / 0	0 / 0	0 / 0	4 / 1
s_{12}	0 / 0	0 / 0	3 / 2	0 / 0
s_{13}	0 / 0	2 / 3	0 / 0	0 / 0
s_{14}	1 / 4	0 / 0	0 / 0	0 / 0

Fig. 1.4. A simple bargaining game

Example 1.1.1 (A simple bargaining game).
Consider a simple bargaining game: Two players have to divide an amount of 5 units and each of them can make one of the following claims—1, 2, 3, or 4. If both claims sum up to 5 then the players get their claims, and otherwise they get nothing. Figure 1.4 shows the payoff matrix of this game, where s_{ik} represents i's claim of k units. An arbitrator suggests a division that gives 3 units to player 1 and 2 units to player 2.

This game has a multitude of Pareto-optimal equilibria, namely, all agreements on splitting the 5 units in some or another way completely. Note that the players' preferences on the set of these agreements are diametral. Nevertheless, they prefer agreement (i.e., an payoff of at least 1) over disagreement (i.e., the payoff 0). So they face a serious equilibrium selection problem, even in view of some degree of common interest.

Schelling (1960) argues that the arbitrator's suggestion makes the respective division the focal one. Of course, there may be other clues that make one division focal, for example, fairness considerations or peculiarities in the representation of the bargaining problem (Shubik 1994). Since there is no other way to reach a favorable outcome, the players should do their part in the focal division given by the arbitrator's suggestion.

However, this is not a consequence of the symmetry-breaking property of the focal division: Without the arbitrator's suggestion, in a sense—to be made precise in Subsection 2.3.5 (pp. 19), both players and their claiming the same amounts—s_{1k} and s_{2k}—are symmetric. Therefore, one could argue that s_{1k} and s_{2k} should be chosen with the same probabilities. But this is not compatible with the suggested division, i.e., player 1 takes action s_{13} and player 2 takes action s_{22}. Since s_{13} is to be taken for sure—probability 1, s_{23} also gets probability 1. According to the arbitrator's suggestion, the same holds for s_{22}. Hence, the sum of probabilities of player 2's strategies s_{21} and s_{23} exceeds 1—a contradiction.

In a sense, the arbitrator's suggestion destroys the symmetry of the players (see Subsection 2.5.6, pp. 50). Now, one of them—player 1—gets a bigger share than the other. Therefore, the strategies are not symmetric, and the arbitrator's suggestion is no longer excluded. But so are the other equilibria that represent agreements, and the players are left with the original selection problem. Hence, the destruction of symmetries does not explain the players' adherence to the focal division.

Originally, Schelling introduced focal points in order to explain efficient bargaining outcomes. As argued above, this explanation cannot be founded on the symmetry breaking capacities of focal points. Hence, it is out of the scope of our theory to be introduced in this thesis. Nevertheless, as a growing body of literature reveals (see e.g. Roth 1985, Mehta, Starmer & Sugden 1991, Binmore, Swierzbinski, Hsu & Proulx 1993, Roth 1995), focal points are important in bargaining. Yet, these focal points do not work by breaking symmetries.

1.2 Dynamic Focal Points

So far, static focal points have been considered only, i.e., focal points that were based on the players' fixed descriptions of games. As argued in Section 4.1, static focal points are important components of the dynamic focal points—focal points that are based on descriptions of strategies that arise in the course of playing the game. These descriptions may be based on preceding choices, either the choices at earlier occurrences of the same decision problem or the choices concerning communication, for example a signal to be sent.

Consider a repeated meeting problem: Suppose a man lost his wife in a department store, and, after a long time of search, they finally met at the toy department. Next time they get separated, in a sense, the toy department is the focal meeting point. When they lost each other the first time, no department was distinguished, and they met by chance only. The second time, the toy department stands out as the previous meeting place and therefore becomes the dynamic focal point.

Although dynamic focal points can be modelled in framed strategic games (see Subsections 4.3.3 and 4.3.4), extensive games are the more appropriate type of games for this purpose. As in strategic games, the symmetry of actions, together with the fact that extensive games do not represent the players' description of actions, prevents the formalization of dynamic focal point considerations in extensive games. Therefore, it is desirable to incorporate these descriptions into game theory in a general way, i.e.,

- to extend extensive games by a structure that represents the players' description of a game, and
- to introduce a solution concept that
  - exploits this structure in order to break some of the symmetries in extensive games
  - in accordance with our intuitions about dynamic focal points.

This is what Chapter 4 is trying to accomplish: We transfer the FSG approach to extensive games. I.e., extensive games are extended by frames into framed extensive games (FEGs), and then, a language invariance requirement similar to that for strategic games restricts the set of equilibria considered as a solution. It is demonstrated by help of some examples that dynamic focal points can be explained by FEG approach within a natural and quite general framework.

1.3 Outline

Developing the framed games approach, the second and the fourth chapter constitute the main parts of this thesis. The third chapter prepares the fourth chapter on dynamic focal points by introducing weak isomorphisms of extensive games. In more detail, the plan of this thesis is as follows:

Chapter 2, firstly, shows how traditional game theory deals with the symmetry problem in strategic games (Section 2.3). Then, starting from the impotence of traditional concepts to explain static focal points, recent attempts to incorporate static focal point considerations into game theory are surveyed (Section 2.4). Combining the traditional concepts and these formalizations of static focal points, we introduce a more general and more basic framework in which static focal points can be explained—the framed strategic game (FSG) approach (Section 2.5). Finally, we discuss the limits of the FSG approach as well as its relation to the other focal point theories (Sections 2.5 and 2.6).

While there is an established notion of symmetry in strategic games (see Section 2.3.4), there is no such notion for extensive games. Mainly, the reason for this state of affairs seems to be the more complex nature of extensive games in connection with the resulting ambiguities in interpreting them. Chapter 3 is concerned with the isomorphisms that underlie the symmetry in extensive games. An important ambiguity concerning isomorphisms is the order of moves. In that chapter, we develop and justify the concept of weak isomorphisms of extensive games. These weak isomorphisms respect the order of moves in a loose sense to be specified. Additionally, we compare weak isomorphisms with their strong counterparts from the literature. So, Chapter 3 establishes the symmetries to be broken in Chapter 4.

In Chapter 4, firstly, we show by help of an example that dynamic focal points cannot be explained within extensive games (Subsections 4.3.1 and 4.3.2). In contrast, we are able to support dynamic focal points within the FSG approach (Subsections 4.3.3 and 4.3.4) developed in Chapter 2. But it seems to be more natural and less cumbersome to transfer the FSG approach to extensive games in order to explain dynamic focal points. This gives rise to the introduction of framed extensive games (FEGs) in Section 4.4. Reconsidering the example, it is demonstrated how the FEG approach can be exploited to formalize dynamic focal points (Section 4.5).

2 Static Focal Points in Framed Strategic Games

2.1 Introduction

Traditional game theoretic solution concepts implicitly regard the strategy labels and the player labels used as the game theorist's ones; rational players should not be concerned with these labels. Harsanyi & Selten (1988) express this view most clearly with their symmetry invariance requirement for solutions of games. An exact definition is given in Subsection 2.3.5. However, this requirement has some unfortunate implications: Consider, to take the most striking example, two-player one-shot matching games—both players have the same (finite) option set and both get a prize only if they independently choose the same option. Since the players cannot distinguish the options in terms of payoff alone, symmetry invariance prescribes them to randomize over the option set. Viewed as a prescription, however, this is highly inefficient—it renders the players with payoffs far lower than in the case of coordination.

In fact, observations in real life (e.g., in natural meeting problems) show that people do not blindly randomize over the whole range of options. On the contrary, they are often quite successful in achieving coordination—at least much more successful than one could predict from randomizing guided by the symmetry invariance requirement. In real life, of course, people do not perceive their decision problems only in terms of payoff differences between strategies. Besides the payoff structure, other features of the game are systematically exploited by the players to determine their actions. It matters how a player perceives the game and how he expects the others to describe it. Besides Schelling's early sample, recent experiments (e.g., Rubinstein & Tversky 1993; Mehta, Starmer & Sugden 1994a, 1994b; Bacharach & Bernasconi 1997) support these observations.

In his famous book *The Strategy of Conflict*, Schelling (1960) introduces the notion of *focal points* to denote this systematic use of non-strategic features in strategic contexts, especially in coordination games. He suggests that "[m]ost situations ... provide some clue for coordinating behavior, some focal point for each person's expectation of what the other expects him to expect to be expected to do" (p. 56). These clues are vaguely characterized as prominent or conspicuous, and unique. Of course, they lie in the players description of the strategies. Similarly, Lewis (1969) coins the term of *salience*—uniqueness in some conspicuous respect.

Although there is some common understanding among game theorists that focal points guide human behavior, only recently have attempts been made to formalize the underlying idea of salience (Bacharach 1991, 1993; Sugden 1995; Janssen 1997; Bacharach & Stahl 2000; see Section 2.4). The reason for this seems to be partly that the notion of salience is difficult to formalize. Schelling (1960, p. 58) suggests that "... we are dealing with imagination as much as with logic; and the logic is of a fairly casuistic kind. Poets may do better than logicians at this game, which is more like 'puns and anagrams' than like chess. Logic helps ... but usually not until imagination has selected some clue to work on from among the concrete details of the situation." Clearly, a formalization of salience will have to catch some of this imagination. More precisely, it involves two parts—(a) a formal structure that represents the players' apprehension of the game—imagination—and (b) a mechanism to derive a salient option from this structure—logic.

This chapter is organized as follows: The second section gives two examples of focal points. Harsanyi & Selten's (1988) symmetry invariance requirement is presented and motivated in the third section. In the forth section, other attempts to formalize focal points are surveyed. Either directly or—as argued in Subsection 2.4.3—indirectly, these approaches invoke a principle of insufficient reason by making up perceived-option sets via coverage functions. But these coverage functions can be made up in several ways.

In order to generalize and to unify the earlier approaches to some extent, we introduce *framed strategic games* (FSGs) together with a *language invariance* requirement upon the solutions of these FSGs in the sixth section. Our concept is related to both Harsanyi & Selten's (1988) symmetry invariance requirement and Bacharach's (1991, 1993) multidimensional labelling: Strategic games are extended by so called *frames* that model the players' apprehension of the game via multidimensional strategy labels, giving rise to FSGs. Then, the symmetry invariance requirement is extended to the solutions of these FSGs—language invariance. In the second part of the sixth section and in the final section, we demonstrate that essential features of the other formalizations of focal points can be represented by the FSG approach.

2.2 Examples

Two examples of static focal points frequently will be reconsidered in this chapter:

Example 2.2.1 (Janssen 1997).
Consider a one-shot matching game where two players independently have to choose a physical object from a set consisting of (1) a light ball, (2) a light cube, and (3) a dark ball (see Figure 2.1).

As mentioned above, randomization is the only symmetry invariant strategy combination in the strategic game representation of this decision situa-

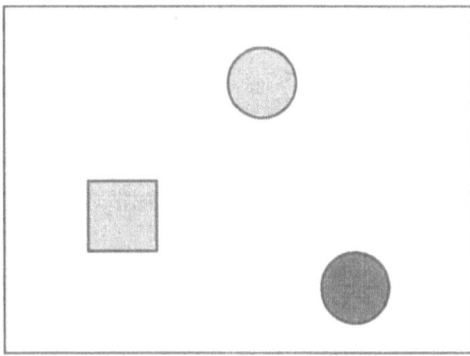

Fig. 2.1. Pure coordination game #3

tion. However, players who realize the color of the objects only can end up much better—the single dark object stands out, and choosing it secures coordination. Analogously, players who are aware of the shape only can achieve coordination by taking the cube for sure. Things are a bit more complicated when the players realize both properties—color and shape. In this case, one could argue that, for example, color as a property is much likelier to come to the mind of the players as a coordinating device than the property of shape. Therefore, even in view of the ambiguity of the single-property prescriptions, the dark ball will be taken.

Consider now the case that both properties and the actual colors and shapes come to mind with the same intensity. Now, the same arguments that lead to take the light cube can be taken to support the choice of the dark ball. Thus, the players may be able to recognize a more subtle kind of salience in this game which enables them to coordinate with certainty. They may realize that both the light cube and the dark ball are unique with respect to one property, whereas the light ball is not unique with respect to any of the properties. Thus, the light ball stands out and should be chosen.

Janssen (1997) formalizes this argument in the spirit of Bacharach's variable frame theory of focal points (see Subsection 2.4.2): Assume that the players realize at least one of the properties color or shape. So each player has three types—color-(only-)realizer, shape-(only-)realizer, and color-shape-realizer. Further, we assume that a color-realizer (shape-realizer) is not aware of the other two types while a color-shape-realizer is aware of all three types. Therefore, the best what color-realizers can do is to take the color-distinguished object—the dark ball, and the best what shape-realizers can do is to take the white cube. Let v_C (v_S) be the probability that a player realizes the objects' color (shape) only.

Assume $v_C + v_S < 1$. Subfigure 2.2 (a) gives the payoff of player 1's color-shape-realizer type depending on the actions of both players' color-shape-realizer types. As indicated above, we assume that both players' color-realizer

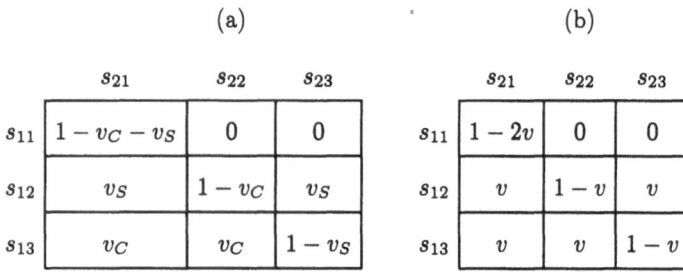

Fig. 2.2. Two payoff matrices

types take the dark ball and that both players' shape-realizer types take the light cube. The first subscript of the strategy-indices s_{ij} indicates the player, $i \in \{1,2\}$, and the second subscript j indicates the chosen object numbered as above. The following example demonstrates how the matrix entries are obtained: Supposing that both color-shape-realizer types take the white cube (matrix entry in the center), player 1's color-shape-realizer type does not coordinate only if player 2 is of the color-realizer type, i.e., he coordinates with probability $1 - v_C$, hence the payoff $1 - v_C$.

If, for example, $v_C > v_S$ which can interpreted as the property color being likelier to come to mind of the players, then both players' taking the dark ball is payoff dominant. Consider now the case $v_C = v_S = v > 0$ the payoffs of which are given by Subfigure 2.2 (b). In some sense (see pp. 19), the players' second and third strategies are symmetric now; and it could be argued that they should be chosen with the same probabilities. For $v < \frac{1}{4}$, both players taking the light ball is payoff dominant. To some extent, this supports the above argument for choosing the light ball.

For $v_C = v_S = v = 0$, i.e. the players realize both properties for sure, however, all strategies are symmetric, and Janssen suggests randomization. An endogenization of the kind of salience as described above seems to be desirable even for this case.

Example 2.2.2 (Bacharach 1993).
Consider a matching game where the players have to choose among twenty blocks that are identical except for that 2 are red and 18 are yellow, and, not so obvious, one of the yellow blocks has a wavy grain of wood whereas the rest have a straight one.

Bacharach (1993) suggests that its solution depends on how probable it is that a player realizes the differences in the grain of wood. Picking a red block is prescribed if this probability is less than $\frac{1}{2}$, and taking the wavy yellow block is prescribed if it is greater than $\frac{1}{2}$. Again, this is in line with intuition—the likelier it is that the other player realizes the grain pattern the likelier it is for a player that realizes grain patterns to take the yellow block with the wavy grain of wood.

2.3 Harsanyi and Selten's Symmetry Invariance Requirement

2.3.1 Introduction

Multi-person decision problems are formalized by game theorists as games—mathematical structures that represent the players, the players' strategies, and the players' preferences over outcomes. Preferences over outcomes are usually represented as von Neumann-Morgenstern utility indices (payoffs) assigned to the (representations of) strategy combinations that lead to the respective outcomes.

In order to identify and to distinguish them, game theorists assign names (indices) to the players and the strategies. Without loss of generality, there are several ways open to the game theorist (but not necessarily to the players) to do this indexing. For example, game theorists could refer to physical properties of the strategies, i.e., the strategy-indices in a game are codes for these properties. Consider the following example:

Example 2.3.1.
Two decision-makers—Ada and Onno—are each given a basket containing a white ball and a black cube. Independently, they have take out one of the objects. If both choose the same object then they win a prize, otherwise they get nothing.

In this case, one game theorist might code the players' first name and the color of the objects to distinguish the strategies while another codes the players' sex and the objects' shape (see pp. 17).

In contrast, Sugden (1995) departs from an 'existential game': a description of the decision problem that only makes existence statements with regard to the number of players and strategies and with regard to the interrelation of strategy combinations and outcomes. In such a description, there is no need for player indices and strategy indices. The above example could be described existentially in the following way: There are two players; each of them has two strategies. For each of the strategies of one player there is exactly one strategy of the other player such that the combination of these strategies results in a prize for both of them while a combination with the other strategy of that player gives both of them nothing. Note that the players and their strategies cannot be distinguished within this framework. In order to make the description less cumbersome, we are free, in doing proofs in first-order logic, to begin by assigning 'arbitrary indices' to players and strategies, a move sometimes called 'existential specification'. In logic, the set of admissable indices is specified in setting up the formal language.

In any case, it seems to be quite natural to require that the game theorists' recommendation for the players or their predicted course of action should not depend on the choice of the representation of the decision problem. Especially, they should be independent of the (arbitrary) player indices and strategy indices.

2.3.2 Strategic Games

Usually, the elements of strategic games are called 'players' and 'strategies', but quite often, the same terms are used to name their referents in real life (see e.g. Harsanyi & Selten 1988, pp. 70). And usually, this does no harm, but it does in the present context, which is all about alternative ways of referring. So in this chapter, we carefully distinguish between players and strategies as real-life entities on the one hand, and their representatives in the game—player-indices and strategy-indices—on the other.

A strategic game G is a tuple $(I, (S_i)_{i \in I}, (u_i)_{i \in I})$, where I denotes the non-empty set of player-indices, S_i the non-empty set of player-index i's pure strategy-indices, and u_i the payoff function $u_i : S \to \mathbb{R}$ where $S = \bigtimes_{i \in I} S_i$. Player-index i's strategy-index in a pure-strategy-index combination $s \in S$ is denoted by s_i. Furthermore, Σ_i denotes player-index i's set of (mixed) strategy-indices σ_i where $\sigma_i(s_i)$ is the probability of player-index i's pure strategy-index s_i in σ_i. Sometimes, s_i stands for the mixed strategy-index σ_i in which s_i gets the probability 1. The set $\bigtimes_{i \in I} \Sigma_i$ of the strategy-index combinations σ is denoted Σ. A strategic game is called finite if I and all S_i are finite; $\mathcal{G}$ denotes the class of finite strategic games.

Many of the applications considered in this work are related to two-player one-shot matching games:

Definition 2.3.1 (Two-player one-shot matching games).
For a given $n \in \mathbb{N}_+$, the matching game M_n is a strategic game $(I, (S_i)_{i \in I}, (u_i)_{i \in I})$ with $I = \{1, 2\}$, $S_i = \{s_{i1}, \ldots, s_{in}\}$ for $i \in I$, $u_i(s_{1j}, s_{2k}) = 1$ if $j = k$, and $u_i(s_{1j}, s_{2k}) = 0$ if $j \neq k$ for all $j, k \in \{1, \ldots, n\}$.

The payoff functions $(u_i)_{i \in I}$ are extended to mixed strategy-index combinations in the usual way: We set

$$u_i(\sigma) := \sum_{s \in S} u_i(s) \prod_{i' \in I} \sigma_{i'}(s_{i'}) \tag{2.1}$$

for all $i \in I$ and $\sigma \in \Sigma$. Hence, the payoffs are linear in one player's strategy-index probabilities, i.e.,

$$u_i(\sigma) = \sum_{s_{i'} \in S_{i'}} \sigma_{i'}(s_{i'}) u_i(s_{i'}, \sigma_{-i'}) \tag{2.2}$$

holds for all $i, i' \in I$ and $\sigma \in \Sigma$.

For any non-empty $I' \subseteq I$, $S_{I'} := \bigtimes_{i \in I'} S_i$ denotes the set of pure I'-subcombinations $s_{I'}$, and $\Sigma_{I'} := \bigtimes_{i \in I'} \Sigma_i$ denotes the set of (mixed) I'-subcombinations $\sigma_{I'}$. Sometimes, the sets $I \setminus \{i\}$ and $I \setminus I'$ are abbreviated as

$-i$ or $-I'$ respectively. For any non-empty and disjoint $I', I'' \subseteq I$, $(s_{I'}, s_{I''})$ and $(\sigma_{I'}, \sigma_{I''})$ denote the $I' \cup I''$-subcombinations in which the players follow $s_{I'}$ and $s_{I''}$ or $\sigma_{I'}$ and $\sigma_{I''}$ respectively.

Definition 2.3.2 (Centroids).
For any $i \in I$ and any non-empty $S_i' \subseteq S_i$, $c_i(S_i') \in \Sigma_i$ denotes the (mixed) strategy-index that assigns probability $|S_i'|^{-1}$ to all pure strategy-indices contained in S_i' and probability 0 to the other pure strategy-indices of $S_i \backslash S_i'$—the centroid of S_i'. We abbreviate $c_i := c_i(S_i)$ for all $i \in I$.
For any non-empty $I' \subseteq I$, $c_{I'} \subseteq \Sigma_{I'}$ denotes the subcombination $(c_i)_{i \in I'}$. We abbreviate $c := c_I$.

The superscript 0 at sets of (mixed) strategy-indices, strategy-index combinations, or subcombinations indicates that all pure strategy-indices involved get non-zero probabilities. I.e., for all $i \in I$, $\Sigma_i \supset \Sigma_i^0 := \{\sigma_i | \forall s_i \in S_i : \sigma_i(s_i) > 0\}$; for any non-empty $I' \subseteq I$, $\Sigma_{I'}^0 := \times_{i \in I'} \Sigma_i^0$; and $\Sigma^0 := \Sigma_I^0$.

Harsanyi & Selten (1988) employ the following strong version of dominance concerning strategy-index combinations:

Definition 2.3.3.
A strategy-index combination σ^* dominates the strategy-index combination σ^{**} if $u_i(\sigma^*) > u_i(\sigma^{**})$ for all $i \in I$. A strategy-index combination is called undominated if there exists no other strategy-index combination that dominates it.

2.3.3 Solution Concepts

Originally, Harsanyi & Selten (1988) consider solution functions only, i.e., one-point solution concepts. Since nearly all equilibrium concepts are not one-point ones, here, a more general version is considered.

Definition 2.3.4 (Solution concepts for strategic games).
A solution concept L for $\mathcal{G}$ assigns a set of strategy-index combinations $L(G) \subseteq \Sigma$—the solutions—to every finite strategic game G.

The most important solution concept in non-cooperative game theory probably is Nash's (1950) equilibrium concept (Myerson 1991, p. 105).

Definition 2.3.5 (Nash equilibrium).
A (Nash) equilibrium of a strategic game G is a strategy-index combination σ^* such that $u_i(\sigma^*) = \max_{\sigma_i \in \Sigma_i} u_i(\sigma_i, \sigma_{-i}^*)$ for all $i \in I$.

Applying the Kakutani fixed point theorem (see Border 1989, pp. 71) to the best-reply correspondences, Nash (1950) shows that any finite strategic game has a Nash equilibrium. In contrast, Nash's (1951) second proof rests upon the more elementary Brouwer fixed point theorem (see Border 1989, pp. 28). Since the technique used in this proof is also employed to prove the existence of symmetry invariant equilibria and symmetry invariant proper equilibria (see Section 2.3.5), this version of the proof is presented in the following. Note that similar techniques are employed in order to establish the existence of equilibria in exchange economies (see e.g. Hildenbrand & Kirman 1988, pp. 106).

Theorem 2.3.1.
Any finite strategic game has a Nash equilibrium.

Proof. The idea of the proof is the following: Firstly, a continuous mapping of Σ onto itself is defined, which has a fixed point according to the Brouwer fixed point theorem. Then, it is shown that any such fixed point is a Nash equilibrium.

Let $G = (I, (S_i)_{i \in I}, (u_i)_{i \in I})$ be a finite strategic game. For all $i \in I$ and $s_i \in S_i$, the mapping $\phi_{s_i} : \Sigma \to \mathbb{R}$ is defined by

$$\phi_{s_i}(\sigma) := \max(0, u_i(s_i, \sigma_{-i}) - u_i(\sigma)) \tag{2.3}$$

for all $\sigma \in \Sigma$. Further, let the mapping $T = (T_i)_{i \in I} : \Sigma \to \Sigma$ be such that

$$T_i(\sigma)(s_i) := \frac{\sigma_i(s_i) + \phi_{s_i}(\sigma)}{1 + \displaystyle\sum_{s_i' \in S_i} \phi_{s_i'}(\sigma)} \tag{2.4}$$

for all $i \in I$, $s_i \in S_i$, and $\sigma \in \Sigma$. Obviously, T is continuous.

Since Σ is compact, convex, and non-empty, and T is continuous, T has a fixed point σ^* according to the Brouwer fixed point theorem. By (2.2), for all $i \in I$, there is some $s_i'' \in S_i$ such that $\sigma_i^*(s_i'') > 0$ and $\phi_{s_i''}(\sigma^*) = 0$. As σ^* is a fixed point of T, by (2.4), we have

$$\sigma_i^*(s_i'') = \frac{\sigma_i^*(s_i'') + \phi_{s_i''}(\sigma^*)}{1 + \displaystyle\sum_{s_i' \in S_i} \phi_{s_i'}(\sigma^*)}$$

and therefore $\sum_{s_i' \in S_i} \phi_{s_i'}(\sigma^*) = 0$, i.e., $\max(0, u_i(s_i, \sigma_{-i}^*) - u_i(\sigma^*)) = 0$ for all $s_i \in S_i$. But this means that no player can improve his payoffs by unilaterally deviating from σ^*—σ^* is a Nash equilibrium. $\square$

2.3.4 Invariance with Respect to Isomorphisms

The labels that players may attach to strategies are not formalized in strategic games. Denotations as player-index 1 or strategy-index s_{12} are—as discussed above—the game theorist's arbitrary labels and not the players'. Generally, two game theorists formalizing the same decision problem are likely to use different strategy-indices and player-indices, and—since independent positive affine transformations of the payoffs do not alter the represented preferences (see e.g. Kreps 1988, pp. 46, 102, 136)—different payoff scales.

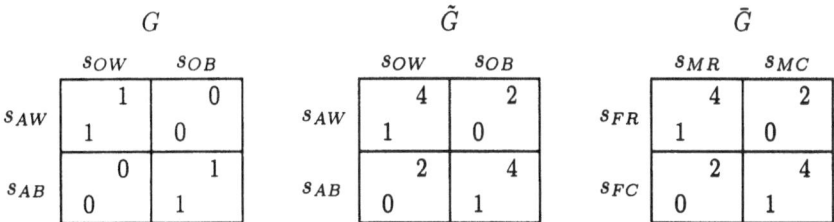

Fig. 2.3. Isomorphic strategic games

An Example. Consider Example 2.3.1. Coding the players' first name and the objects' color game theorist T might set up the strategic game G with $I = \{A, O\}$, $S_A = \{s_{AW}, s_{AB}\}$, $S_O = \{s_{OW}, s_{OB}\}$, and payoffs assigned as in Figure 2.3 where the lower left ones refer to the row player-index and the upper right ones to the column player-index. In G, for example, the player-index A (O) is the code for the name 'Ada' ('Onno') and the strategy-index s_{AB} (s_{OW}) the code for Ada's strategy of taking the black cube (Onno's taking the white ball).

Another game theorist $\bar{T}$ who codes the players' sex and the objects' shape and who uses other utility scales models this decision problem by the strategic game $\bar{G}$ with $\bar{I} = \{M, F\}$, $S_M = \{s_{MR}, s_{MC}\}$, $S_F = \{s_{FR}, s_{FC}\}$, and payoffs assigned as in Figure 2.3 where the player-index F (M) stands for the female Ada (male Onno) and the subscript R (C) for taking the round (cubic) object.

Isomorphisms. In some sense, the games G and $\bar{G}$ in Figure 2.3 are similar; $\bar{G}$ can be obtained from G by re-indexing players and strategies and by re-scaling the payoffs. In order to formalize this idea, Harsanyi & Selten (1988) introduce isomorphisms of strategic games:

Definition 2.3.6 (Isomorphisms of strategic games).
An isomorphism from $(I, (S_i)_{i \in I}, (u_i)_{i \in I}) \in \mathcal{G}$ to $(\bar{I}, (\bar{S}_{\bar{i}})_{\bar{i} \in \bar{I}}, (\bar{u}_{\bar{i}})_{\bar{i} \in \bar{I}}) \in \mathcal{G}$ is a system of bijective mappings $\mathbf{f} = (\pi, (r_i)_{i \in I})$, $\pi : I \to \bar{I}$ and $r_i : S_i \to \bar{S}_{\pi(i)}$, with the following property: For all $i \in I$, there are $\alpha_i, \beta_i \in \mathbb{R}$, $\alpha_i > 0$ such that

$$\bar{u}_{\pi(i)} (f(s)) = \alpha_i u_i(s) + \beta_i \qquad (2.5)$$

holds for all $s \in S$ where $f = (f_{\bar{i}})_{\bar{i} \in \bar{I}}$ denotes the mapping $f : S \to \bar{S}$, $s \mapsto f(s)$ with

$$f_{\pi(i)}(s) := r_i(s_i) \qquad (2.6)$$

for all $i \in I$ and $s \in S$.

Obviously, a system of identity mappings is an isomorphism from a strategic game to itself. Since all mappings involved are bijective, and compositions and inverses of positive affine transformations are positive affine, inverses and compositions of isomorphisms are also isomorphisms. Thus, isomorphisms constitute an equivalence relation on $\mathcal{G}$. Two strategic games are called isomorphic if they can be mapped onto each other by an isomorphism. Note that the notions of isomorphisms of strategic games and of isomorphic strategic games are similar to the respective notions for one-person decision problems (games against nature) in Arrow & Hurwicz (1972).

Consider the games in Figure 2.3. Positive affine transformations of the payoffs with $\alpha_A = 1$, $\beta_A = 0$, $\alpha_O = 2$, and $\beta_O = 2$ transform G into $\tilde{G}$. Thereafter, re-indexing $\tilde{G}$ by $\pi(A) = F$, $\pi(O) = M$, $r_A(s_{AW}) = s_{FR}$, $r_A(s_{AB}) = s_{FC}$, $r_O(s_{OW}) = s_{MR}$, and $r_O(s_{OB}) = s_{MC}$ transforms $\tilde{G}$ into $\bar{G}$. Thus, $(\pi, (r_A, r_O))$ is an isomorphism from G to $\bar{G}$; G and $\bar{G}$ are isomorphic. But since G and $\bar{G}$ represent the same decision problem, this is not astonishing.

Invariance. The other way round, isomorphic strategic games can be considered to represent the same decision problem. Looking at (isomorphic) strategic games alone, it is impossible to say whether they represent the same decision problem or not. Therefore, it could be argued that isomorphic strategic games should be assigned the same solution. This way the solution for a decision problem would always be independent of the game theorist's description. Harsanyi & Selten (1988) formalize this idea by requiring invariance with respect to isomorphisms for solution functions for $\mathcal{G}$. Below, the straightforward extension to solution concepts is given. Note that a similar invariance criterion for solutions of decision problems has been proposed by Milnor (1954) and Arrow & Hurwicz (1972): rational choice should be independent of the ordering or labelling of actions and states of nature.

The mapping $f : S \rightarrow \bar{S}$ in Definition 2.3.6 can be extended to Σ by setting

$$f_{\pi(i)}(\sigma)(r_i(s_i)) := \sigma_i(s_i) \qquad (2.7)$$

for all $i \in I$, $s_i \in S_i$, and $\sigma \in \Sigma$. Let $\mathbf{f}$ be an isomorphism between two strategic games. Then, we have

$$\bar{u}_{\pi(i)}(f(\sigma)) = \alpha_i u_i(\sigma) + \beta_i$$

for all $i \in I$ and $s \in S$ where f and all α_i and β_i are determined by $\mathbf{f}$ via (2.6) and (2.5) respectively.

Definition 2.3.7 (Invariance with respect to isomorphisms).
A solution concept L for $\mathcal{G}$ is invariant with respect to isomorphisms if for every isomorphism $\mathbf{f}$ from $G \in \mathcal{G}$ to $\bar{G} \in \mathcal{G}$ we have $f(L(G)) = L(\bar{G})$ for f given by $\mathbf{f}$ via (2.6).

Harsanyi & Selten (1988) show that the Nash equilibrium concept is invariant with respect to isomorphisms. In general, all traditional equilibrium (solution) concepts have this property. This assertion is not proven rigorously. But its plausibility can be seen from the following facts: (a) None of the traditional equilibrium (solution) concepts makes use of player or strategy indices.

(b) All of them are based on the signs of single players' payoff differences—regarding mixed strategy-index combinations—only. Clearly, positive affine transformations as those in the definition of isomorphisms—(2.5)—do not change these signs.

2.3.5 Symmetry Invariance

Symmetries. Automorphisms of a strategic game G are isomorphisms from G onto itself. Harsanyi & Selten (1988) call these automorphisms the symmetries of G. Note that the symmetries considered by Nash (1951) are slightly different—in (2.5), he fixes all α_i at 1 and all β_i at 0.

Invariance. Reconsider Example 2.3.1. Suppose there existed a language Glishen where the meanings of the words are exchanged such that 'male' means 'female' and vice versa, and 'white' means 'black' and vice versa. Nevertheless, a Glishen-speaking game theorist $\tilde{T}$ might represent this decision problem by the strategic game $\bar{G}$ (see Figure 2.3). But in contrast to $\bar{G}$ made up by the English-speaking game theorist $\bar{T}$, for example, the player-index M now represents the Ada instead of Onno and the strategy index s_{FR} represents Onnos's strategy of taking the black cube instead of Ada's strategy of taking white ball. The different meanings of the indices in $\bar{G}$ for $\bar{T}$ and $\tilde{T}$ can be related to another by the following symmetry of $\bar{G}$: $\bar{\pi}(M) = F$, $\bar{\pi}(F) = M$, $r_F(s_{FR}) = s_{MC}$, $r_F(s_{FC}) = s_{MR}$, $r_M(s_{MR}) = s_{FC}$, and $r_M(s_{MC}) = s_{FR}$. For example, $r_M(s_{MC}) = s_{FR}$ can be interpreted as s_{MC} having the same meaning to $\tilde{T}$ as s_{FR} to $\bar{T}$, namely Onno's strategy of taking the black cube.

But the game $\bar{G}$ alone does not allow us to say whether s_{MC} or s_{FR} stands for Onno's strategy of taking the black cube. Since both strategy-indices could represent the same strategy, they should be treated in the same way, i.e., they should be assigned the same probability in a solution. This is formalized by the following definition:

Definition 2.3.8 (Symmetry invariance).
*A strategy-index combination σ is symmetry invariant if $f(\sigma) = \sigma$ for all f given by symmetries **f** of G via (2.6).*

Symmetric Strategy-Indices and Symmetric Player-Indices. Sometimes, it is more convenient and—in a sense—more natural to express symmetry invariance in terms of symmetric strategy-indices. Using symmetries of strategic games, we define symmetric strategy-indices and symmetric player-indices as follows:

Definition 2.3.9 (Symmetric strategy-indices).
Two pure strategy-indices $s_i \in S_i$ and $s_{i'} \in S_{i'}$ in $G = (I, (S_i)_{i \in I}, (u_i)_{i \in I}) \in \mathcal{G}$ are symmetric, if there is a symmetry $\mathbf{f} = (\pi, (r_i)_{i \in I})$ of G such that $r_i(s_i) = s_{i'}$.

Since isomorphisms constitute an equivalence relation on $\mathcal{G}$, the symmetry of pure strategy-indices is also an equivalence relation. Therefore, this definition is equivalent to the seemingly more general one via a chain of mappings $r_{\pi^n(i)} \circ \ldots \circ r_{\pi^2(i)} \circ r_{\pi(i)} \circ r_i$ for some $n \in \mathbb{N}$, where $\circ$ denotes the composition of mappings. Using Definition 2.3.9, one can characterize symmetry invariant strategy-index combinations as follows:

Corollary 2.3.1.
A strategy-index combination σ in $G \in \mathcal{G}$ is symmetry invariant if and only if all symmetric pure strategy-indices get the same probabilities.

As suggested in Section 2.3.5, for example, the strategy-indices s_{MC} and s_{FR} from $\bar{G}$ (see Figure 2.3) are symmetric, and symmetry invariance requires them to be chosen with the same probabilities. In M_n, all (of one player-index') pure strategy-indices are symmetric and therefore receive the same probabilities in the unique symmetry invariant strategy-index combination.

Crawford & Haller (1990) introduce a similar notion of symmetric strategy-indices for two-player strategic games with identical payoffs for both players.

Definition 2.3.10.
Let $G \in \mathcal{G}$ be a strategic game with $I = \{1,2\}$ and $u := u_1 = u_2$. Let Z_1 be some partition of S_1 and let Z_2 be some partition of S_2. A pair (Z_1, Z_2) is called compatible if it satisfies the following conditions:

1. *For all $z_1 \in Z_1$ and $s_1, s_1' \in z_1$, there exists a bijective mapping $r_2 : S_2 \to S_2$ such that $r_2(z_2) = z_2$ holds for all $z_2 \in Z_2$ and $u(s_1, s_2) = u(s_1', r_2(s_2))$ holds for all $s_2 \in S_2$.*
2. *For all $z_2 \in Z_2$ and $s_2, s_2' \in z_2$, there exists a bijective mapping $r_1 : S_1 \to S_1$ such that $r_1(z_1) = z_1$ holds for all $z_1 \in Z_1$ and $u(s_1, s_2) = u(r_1(s_1), s_2')$ holds for all $s_1 \in S_1$.*

For any $i \in I$, the strategy-indices $s_i, s_i' \in S_i$ are called symmetric if there exists a compatible pair (Z_1, Z_2) and some $z_i \in Z_i$ such that $s_i' \in z_i$ and $s_i \in z_i$.

The theorem below shows that our strategy-index symmetry relation implies Crawford & Haller's strategy-index symmetry relation. However, it is not yet clear whether the converse holds.

Theorem 2.3.2.
Let $G \in \mathcal{G}$ be a strategic game with $I = \{1,2\}$ and $u := u_1 = u_2$. If $s_i, s_i' \in S_i$ are symmetric for some $i \in I$ according to Definition 2.3.9, then s_i and s_i' are symmetric according to Definition 2.3.10.

The proof of this theorem is prepared by a lemma.

Lemma 2.3.1.
Let $\mathbf{f} = (\pi, (r_\iota)_{\iota \in I})$ *be a symmetry of* $G \in \mathcal{G}$ *with* $\pi(i') = i'$ *for some* $i' \in I$. *Then, we have* $u_{i'}(f(\sigma)) = u_{i'}(\sigma)$ *for all* $\sigma \in \Sigma$, *where* f *is given by* $\mathbf{f}$ *via* (2.6).

Proof. Let $\mathbf{f} = (\pi, (r_\iota)_{\iota \in I})$ be a symmetry of $G \in \mathcal{G}$ with $\pi(i') = i'$ for some $i' \in I$; and let $u_{i'}^{\max} := \max_{s \in S} u_{i'}(s)$, $s^{\max i'} \in \arg\max_{s \in S} u_{i'}(s)$, $u_{i'}^{\min} := \min_{s \in S} u_{i'}(s)$, and $s^{\min i'} \in \arg\min_{s \in S} u_{i'}(s)$. By (2.5), we then have $f(s^{\max i'}) \in \arg\max_{s \in S} u_{i'}(s)$ and $f(s^{\min i'}) \in \arg\min_{s \in S} u_{i'}(s)$. Hence, by (2.5),

$$u_{i'}^{\max} = \alpha_{i'} u_{i'}^{\max} + \beta_{i'} \tag{2.8}$$

$$u_{i'}^{\min} = \alpha_{i'} u_{i'}^{\min} + \beta_{i'} \tag{2.9}$$

for some $\alpha_{i'}, \beta_{i'} \in \mathbb{R}$, $\alpha_{i'} > 0$. In case of $u_{i'}^{\max} = u_{i'}^{\min}$, nothing is to show. For $u_{i'}^{\max} > u_{i'}^{\min}$, solving the system of equations (2.8), (2.9) gives $\alpha_{i'} = 1$ and $\beta_{i'} = 0$, i.e., $u_{i'}(f(\sigma)) = u_{i'}(\sigma)$ for all $\sigma \in \Sigma$. $\square$

Proof. (Theorem 2.3.2) Let $G \in \mathcal{G}$ be a strategic game with $I = \{1, 2\}$ and $u := u_1 = u_2$, and let $s_1, s_i' \in S_\iota$ be symmetric for some $i \in I$ according to Definition 2.3.9.

For all $i \in I$, let Z_ι be the set of equivalence classes of S_i with regard to the strategy-symmetry relation of Definition 2.3.9. Obviously, $r_i(z_i) = z_\iota$ holds for all $i \in I$ and $z_\iota \in Z_\iota$. By Definition 2.3.9, there exists a symmetry $(\pi, (r_1, r_2))$ of G such that $r_i(s_\iota) = s_i'$. Hence, we also have $\pi = \mathrm{id}$. Obviously, $r_\iota(z_\iota) = z_\iota$ holds for all $i \in I$ and $z_\iota \in Z_\iota$. By Lemma 2.3.1, we have

$$u(s_1, s_{-\iota}) = u(f(s_i, s_{-i})) = u(r_i(s_i), r_{-i}(s_{-i})) = u(s_i', r_{-i}(s_{-i}))$$

for all $s_{-\iota} \in S_{-\iota}$. Thus, (Z_1, Z_2) is a compatible pair, and s_i and s_i' are symmetric according to Definition 2.3.10. $\square$

It is not ruled out that symmetric strategy-indices belong to different player-indices i and i'. In this case, these player-indices also are called symmetric.

Definition 2.3.11 (Symmetric player-indices).
Two player-indices $i, i' \in I$ *in* $G = (I, (S_i)_{i \in I}, (u_i)_{i \in I}) \in \mathcal{G}$ *are symmetric, if there is a symmetry* $\mathbf{f} = (\pi, (r_\iota)_{\iota \in I})$ *of* G *such that* $\pi(i) = i'$.

Note that symmetric strategy-indices (symmetric player-indices) are just those strategy-indices (player-indices) that are assigned the same description in Sugden's existential specifications of strategic games (see Subsection 2.3.1).

	s_{21}	s_{22}
s_{11}	1 2	0 0
s_{12}	0 0	3 1

Fig. 2.4. A game with several symmetry invariant equilibria

Existence of Symmetry Invariant Equilibria. Invariance with respect to isomorphisms has the strong implication that a one-point solution concept has to assign a symmetry invariant strategy-index combination to any strategic game. Nash's (1951) second theorem ensures that every finite strategic game has a symmetry invariant equilibrium. This qualifies the Nash equilibrium to be part of a one-point solution concept. Quite often, of course, the problem of selecting one Nash equilibrium from a multitude of even symmetry invariant equilibria remains. Consider the strategic game in Figure 2.4. The player-indices and all strategy-indices are not symmetric. Therefore, all (mixed) strategy-index combinations—including the three Nash equilibria—are symmetry invariant.

Theorem 2.3.3 (Nash 1951).
Any finite strategic game has a symmetry invariant equilibrium.

Proof. Again, the idea of the proof is the following: Firstly, it is shown that the set of symmetry invariant strategy-index combinations is non-empty, convex, and compact. Secondly, a continuous mapping of this set onto itself is defined. By the Brouwer fixed point theorem, this mapping has a fixed point. Finally, it is shown that the fixed points of this mapping are Nash equilibria.

Let $G = (I, (S_i)_{i \in I}, (u_i)_{i \in I})$ be a finite strategic game. Obviously, c (see Definition 2.3.2, p. 15) is symmetry invariant. Hence, the set of symmetry invariant strategy-index combinations is non-empty. Let σ and σ' be symmetry invariant, let $\mathbf{f}$ be a symmetry of G, and let $f : \Sigma \to \Sigma$ be defined by $\mathbf{f}$ via (2.6). By (2.7), we have

$$f([a\sigma + (1-a)\sigma'])_{\pi(i)}(r_i(s_i)) = [a\sigma + (1-a)\sigma']_i(s_i)$$
$$= [af(\sigma) + (1-a)f(\sigma')]_{\pi(i)}(r_i(s_i))$$
$$= [a\sigma + (1-a)\sigma']_{\pi(i)}(r_i(s_i)),$$

for all $i \in I$, $s_i \in S_i$, and $a \in [0,1]$, i.e., $f([a\sigma + (1-a)\sigma']) = [a\sigma + (1-a)\sigma']$. Hence, the set of symmetry invariant strategy-index combinations is non-empty, convex, and, since f is continuous, compact.

For all $i \in I$ and $s_i \in S_i$, let $\hat{\phi}_{s_i} : \Sigma \to \mathbb{R}$ be defined by

$$\Phi_i = \max_{s \in S} u_i(s) - \min_{s \in S} u_i(s),$$

$$\hat{\phi}_{s_i}(\sigma) = \begin{cases} \dfrac{\phi_{s_i}(\sigma)}{\Phi_i} & : \Phi_i > 0, \\ 0 & : \Phi_i = 0, \end{cases}$$

where $\phi_{s_i} : \Sigma \to \mathbb{R}$ is defined by (2.3). The mapping $\hat{T} = (\hat{T}_i)_{i \in I} : \Sigma \to \Sigma$ as a modification of the mapping T from (2.4) is set such that

$$\hat{T}_i(\sigma)(s_i) := \frac{\sigma_i(s_i) + \hat{\phi}_{s_{ij}}(\sigma)}{1 + \sum\limits_{s_i' \in S_i} \hat{\phi}_{s_i'}(\sigma)} \tag{2.10}$$

for all $i \in I$, $s_i \in S_i$, and $\sigma \in \Sigma$. In view of the difference between Nash's (1951) and Harsanyi & Selten's (1988) symmetries (see pp. 19), our modification of T is necessary in order to ensure that $\hat{T}$ is a mapping of the set of symmetry invariant strategy-index combinations onto itself.

By (2.5), we have $\hat{\phi}_{r_i(s_i)}(f(\sigma)) = \hat{\phi}_{s_i}(\sigma)$ for all $i \in I$, $s_i \in S_i$, and $\sigma \in \Sigma$. Hence, we have

$$f_{\pi(i)}(\hat{T}(\sigma))(r_i(s_i)) = \hat{T}_i(\sigma)(s_i) = \frac{\sigma_i(s_i) + \hat{\phi}_{s_i}(\sigma)}{1 + \sum\limits_{s_i' \in S_i} \hat{\phi}_{s_i'}(\sigma)}$$

$$= \frac{\sigma_{\pi(i)}(r_i(s_i)) + \hat{\phi}_{r_i(s_i)}(f(\sigma))}{1 + \sum\limits_{s_i' \in S_i} \hat{\phi}_{r_i(s_i')}(f(\sigma))}$$

$$= \frac{\sigma_{\pi(i)}(r_i(s_i)) + \hat{\phi}_{r_i(s_i)}(\sigma)}{1 + \sum\limits_{s_i' \in S_i} \hat{\phi}_{r_i(s_i')}(\sigma)}$$

$$= \hat{T}_{\pi(i)}(\sigma)(r_i(s_i))$$

I.e., $f(\hat{T}(\sigma)) = \hat{T}(\sigma)$ holds for all $\sigma \in \Sigma$—$\hat{T}$ is a mapping of the set of symmetry invariant strategy-index combinations onto itself which has a fixed point σ^* according to the Brouwer fixed point theorem.

For $\Phi_i > 0$, one can argue along the lines of the proof of Theorem 2.3.1 that σ_i^* is a best reply to σ_{-i}^*. In case of $\Phi_i = 0$, all $\sigma_i \in \Sigma_i$ are best replies to all $\sigma_{-i} \in \Sigma_{-i}$. Hence, σ_i^* is best reply to σ_{-i}^*. Since this holds for all i, σ^* is a Nash equilibrium. $\square$

Since the mappings f induced by symmetries via (2.6) are continuous, applying Theorem 2.3.3 to the definition of perfect equilibria (Selten 1975) gives the following corollary.

Corollary 2.3.2.
The concept of perfect equilibria is invariant with respect to isomorphisms.

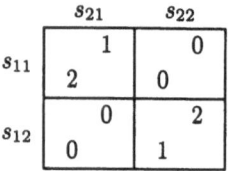

Fig. 2.5. Battle of the Sexes

A similar application within the usual existence proofs for perfect equilibria establishes the existence of symmetry invariant perfect equilibria.

Corollary 2.3.3.
Any finite strategic game has a symmetry invariant perfect equilibrium.

Additionally, combining the techniques used in Nash's (1951, Theorem 2) and Myerson's (1978, Theorem 2) proof, one can show that symmetry invariant proper equilibria always exist. In contrast, the same does not hold for the set based concepts of persistent equilibria (Kalai & Samet 1984) and stable equilibria (Kohlberg & Mertens 1986). Consider the Battle-of-the-Sexes game (see Figure 2.5). Its unique symmetry invariant equilibrium—the mixed one—is neither persistent nor stable.

2.4 Rationalizing Static Focal Points

In view of the empirical observations in matching games, symmetry invariance seems to be too strong a requirement upon the solutions of strategic games. But it reflects the fact that players cannot make any use of indices. Again, this is justified *if* the strategy and player indices are considered to be those of the game theorist. But then, focal points as symmetry-breaking devices cannot be supported within the framework traditional game theory.

This section surveys approaches to formalize this kind of focal points. The first subsection presents Gauthier's (1975) attempt to rationalize the choice of a salient option. Bacharach's (1991, 1993) variable frame theory (VFT) and Janssen's (1997) extension are presented in the second subsection. The third subsection deals with Sugden's (1995) label procedures. The principle of coordination which is employed in all these approaches is discussed in the final subsection.

2.4.1 Rationalizing the Choice of a Salient Option

The first attempt to rationalize the choice of a salient option was made by Gauthier (1975). He suggests that the decision process of players facing a matching game takes two steps: (a) The players decide to reconceive the game. In order to do this, the players first have to conceive one of the options as the

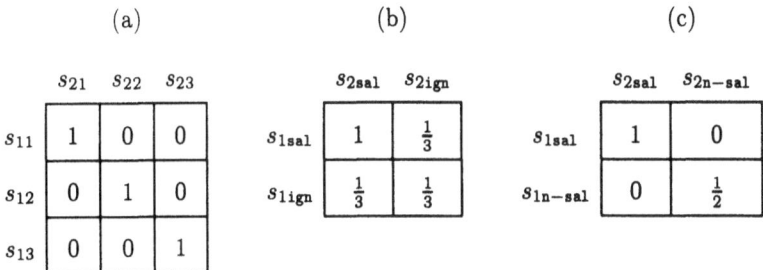

Fig. 2.6. Reconceiving a strategic game by salience

commonly salient one. Next, the players restrict themselves to considering just two actions—either to choose the salient option or to ignore salience and to randomize. (b) Gauthier then invokes a so called *principle of coordination* according to which the players choose that option from the restricted option set that results in a unique undominated equilibrium.

Consider the game M_3 the payoff matrix of which is given in Subfigure 2.6 (a). Suppose both players' first strategies are salient. Reconceiving the game gives the game in Subfigure 2.6 (b), where s_{isal} refers to player i's choosing the salient option and s_{iign} refers to randomization over all options—ignoring salience. Now, taking the salient option is payoff dominant and is therefore selected by the principle of coordination.

Provis (1977), Gilbert (1989), and Miller (1991) object that instead of considering the ignorance of salience as the players' second option after reconceiving the game, one rather should consider picking a non-salient option. For M_3, this gives the reduced game in Subfigure 2.6 (c), where s_{in-sal} refers to player i's picking a non-salient option. Again, the principle of coordination selects the salient option. More general, in M_n, the choice of the salient option can be supported for both ways of reconceiving a game as long as $n > 2$. For $n = 2$, the first way of reconceiving the game supports the choice of the salient option, while the second one does not. Following Provis, Gilbert, and Miller, one could argue that the second way is more appropriate (see pp. 48).

As Goyal & Janssen (1996) remark, this kind of rationalization of choosing a salient option is to be distinguished from the formalization of salience itself. However, this two-step process is an important step towards a rationalization of salience itself (Bacharach 1991), even in view of Gilbert's (1989) criticism. For this purpose, of course, alternative ways of restricting the perceived-option set will have to be used. The coverage functions of the next subsection and the language invariance requirement of Subsection 2.5.2 suggest such alternatives.

2.4.2 The Variable Frame Theory of Focal Points

The basic ideas of Bacharach's (1991, 1993) variable frame theory (VFT) are
(a) that the players are given descriptions of their strategies with respect to
a number of attributes (e.g., shape or color of an object to choose), (b) that
the set of attributes a player actually is aware of (his repertoire) restricts
his perceived options by applying a principle of insufficient reason, (c) that
there is a commonly known probability distribution over these repertoires
which formalizes the idea of prominence (Schelling 1960) or conspicuous-
ness (Lewis 1969), and (d) that a player cannot think of players who perceive
attributes that he does not realize. Together with the principle of payoff dominance, VFT prescribes intuitive solutions to some matching games. In many
cases, however, VFT does not lead to a (unique) solution. In order to remedy
this ambiguity, Janssen (1997) introduces a general solution concept for a similar class of games that is based on a notion of symmetry between perceived
options and an iterative reduction mechanism on the sets of perceived options
formed by repertoires. After reduction, a *principle of individual team member rationality* (an 'individualistic' payoff dominance criterion) determines
the solution for which Janssen proves generic uniqueness, equilibrium, and
efficiency properties. Bacharach & Stahl (2000) present a boundedly rational
version of VFT.

Variable Universe Matching Games (VUMGs). We consider *variable
universe matching games* (VUMGs) that are based on Bacharach's (1991,
1993) variable universe games with some modifications made by Bacharach
& Stahl (2000):

VUMGs $(M_n, \mathfrak{A}, (\ell_{ia})_{(i,a) \in I \times \mathfrak{A}}, v)$ are two-player one-shot matching games
M_n (see Definition 2.3.1, p. 14) in which the players realize several properties
of the strategies. For each property there is an attribute $a \in \mathfrak{A}$ containing
labels that the players can use to describe the strategies with respect to this
property, for example, the attribute color comprising color labels such as
red, green and blue. It is assumed that the attributes are pairwise disjoint.
For all attributes $a \in \mathfrak{A}$ and both players, the label functions $\ell_{ia} : S_i \to a$
assign a label to each strategy with respect to a certain attribute. Without
loss of generality, we set $\ell_{1a}(s_{1j}) = \ell_{2a}(s_{2j}) \in a$, where $\ell_{ia}(s_{ij})$ is the label that
describes the jth strategy of player i with respect to attribute a, i.e., strategies
with the same number are assigned the same labels. The set of actually
realized attributes, a subset of $\mathfrak{A}$, is called a player's repertoire $r \in \mathbf{R} := 2^{\mathfrak{A}}$;
subsets of r are called its subrepertoires. Alternatively, Bacharach & Stahl
(2000) call this set a frame.

Example 2.4.1.
Consider the game from Example 2.2.1 (p. 10). We have $\mathfrak{A} = \{\texttt{color}, \texttt{shape}\}$,
color = $\{\texttt{light}, \texttt{dark}\}$, shape = $\{\texttt{round}, \texttt{cubic}\}$, and $(\ell_{ia})_{(i,a) \in I \times \mathfrak{A}}$ as given
by Figure 2.7 for all $i \in I$. The repertoires then are $\emptyset$, $\{\texttt{color}\}$, $\{\texttt{shape}\}$, and
$\mathfrak{A}$ itself, where the empty repertoire stands for players who realize neither

s_{ij}		s_{i1}	s_{i2}	s_{i3}
a	$\ell_{ia}(s_{ij})$			
color		light	light	dark
shape		round	cubic	round

Fig. 2.7. A label function

the color nor the shape of the objects; the repertoire {color} ({shape}) represents players who realize the colors (shapes) of the objects only; and players who are aware of both properties are characterized by the repertoire $\mathfrak{A}$.

The distribution of repertoires v in the population of players is given exogenously. In a sense, it formalizes the idea of prominence (Schelling 1960) or conspicuousness (Lewis 1969). The share of repertoire r is denoted $v(r)$. The conditional probability $v(r'|r)$ that a player with repertoire r expects the other player to have repertoire r' can be derived from v by

$$v(r'|r) := \begin{cases} \frac{v(r')}{\sum_{r'' \subseteq r} v(r'')} & : r' \subseteq r, \sum_{r'' \subseteq r} v(r'') \neq 0, \\ |r|^{-1} & : r' \subseteq r, \sum_{r'' \subseteq r} v(r'') = 0, \\ \text{not defined} & : r' \not\subseteq r. \end{cases} \quad (2.11)$$

Bacharach & Stahl (2000) justify this derivation as an implication of rational expectations.

The denominator in the first line represents the share of players who have the repertoire r or some of its subrepertoires, i.e., it represents the share of players that a player with repertoire r realizes. This way, the whole fraction, indeed, gives the conditional probability as indicated. The second line concerns repertoires that get zero probability, i.e., repertoires that cannot be found in the population of players. It ensures that the conditional probability is defined for all subrepertoires, even in case that subrepertoire players do not exist. As a purely technical assumption, all subrepertoires get the same probability. The third line means that a player cannot think of attributes that are not contained in his repertoire. Therefore, players with different repertoires face different games. In contrast to their interpretation based on *objective* population shares, the conditional probabilities $v(r'|r)$ can be viewed as the players' *subjective* expectations concerning the other player's repertoire.

A crucial assumption of Bacharach's (1993) variable frame theory is that the distribution of the repertoires is the same for all players. This excludes for example the possibility that two players with the same repertoire have different expectations concerning subrepertoire players. This seemingly restrictive assumption, however, can be justified as an expression of a shared culture. Another justification considers this distribution as an commonly known public event (Bacharach 1998).

Coverage Functions and Principle of Insufficient Reason. Different repertoires may give rise to different perceived-option sets which are assigned to both players i with regard to all repertoires by coverage functions that are—in abuse of notation—denoted S_i. The coverage function $S_i : \mathbf{R} \rightrightarrows \Sigma_i$ assigns the perceived-option set $S_i(r) \subset \Sigma_i$ to player i with regard to the repertoire $r \in \mathbf{R}$, where $S_i(r)$ contains the (pure) options that player i considers—given his description of the options based on the attributes contained in his repertoire r. In a sense, these coverage functions are expressions of the principle of insufficient reason—strategies the labelling of which is sufficiently similar are to be taken with the same probabilities by a perceived option. In addition, they can be viewed as a generalization of Gauthier's (1975) way of reconceiving the game.

For example, Bacharach & Bernasconi (1997) specify the following perceived-option set for a player with repertoire r. Each label λ in one of the attributes $\mathfrak{a}$ of the repertoire r that actually is attributed to one of the pure strategies contributes one perceived option to this set—picking a strategy with this label, i.e., the randomization over all pure strategies that are characterized by this label (see Morgenbesser & Ullmann-Margalit 1977). In addition, this set contains the randomization over the whole option set—c_i (see Definition 2.3.2, p. 15). More formally, in abuse of notation, let $S_i(\mathfrak{a}, \lambda) \subseteq S_i$ denote the possibly empty set of player i's pure strategy-indices that are assigned the label λ with regard to the attribute $\mathfrak{a}$. Then, we have

$$S_i(r) := \{c_i(S_i(\mathfrak{a}, \lambda)) | \mathfrak{a} \in r, \lambda \in \mathfrak{a} : S_i(\mathfrak{a}, \lambda) \neq \emptyset\} \cup \{c_i\} \qquad (2.12)$$

for all $i \in I$ and $r \in \mathbf{R}$.

Example 2.4.2.
Reconsider Example 2.4.1. For players with the repertoire $\{\texttt{color}, \texttt{shape}\}$, we have five perceived options: (a) overall randomization—c_i, (b) picking a light object—$c_i(S_i(\texttt{color}, \texttt{light}))$, (c) choosing the dark ball—$c_i(S_i(\texttt{color}, \texttt{dark}))$, (d) picking a ball—$c_i(S_i(\texttt{shape}, \texttt{round}))$, and (e) choosing the light cube—$c_i(S_i(\texttt{shape}, \texttt{cubic}))$.

This coverage function represents players who cannot combine different attributes in order to make up the perceived-option set. Although all objects are distinguished by color and shape together, choosing the light ball is no perceived option. Therefore, coverage functions that cover the combination of attributes seem to be more appropriate, at least in the case of few attributes only. Of course, it is an empirical matter to which extent players are capable of combining attributes. Such combining coverage functions are suggested by Janssen (1997) and Bacharach & Stahl (2000).

Janssen (1997) introduces coverage functions that rest upon partitions of the players' pure-strategy sets that are based on the labelling of the respective repertoires. For all repertoires $r \in \mathbf{R}$, let $\sim_{i,r}$ be an equivalence relation on S_i as follows: For all $s_i, s_i' \in S_i$, we have $s_i \sim_{i,r} s_i'$ iff $\ell_{i\mathfrak{a}}(s_i) = \ell_{i\mathfrak{a}}(s_i')$ for all

$\mathfrak{a} \in r$. I.e., two strategies are equivalent with regard to a repertoire r if they are labelled in the same way with regard to this repertoire's attributes. Let $[s_i]_r$ denote the equivalence class of s_i concerning $\sim_{i,r}$. Then, we define

$$S_i(r) := \{c_i([s_i]_r)|s_i \in S_i\} \tag{2.13}$$

for all $i \in I$ and $r \in \mathbf{R}$. Whereas the coverage function in (2.12) does not combine the attributes at all, this one combines them in the strictest possible sense—strategy-indices with identical label vectors only form a perceived option.

This way, as indicated in Section 2.2, some subtle kinds of salience cannot be explained. Reconsider Example 2.4.1. For players with the repertoire {color, shape}, the perceived-option set is identical to set of pure strategies: (a) choosing the light ball—$s_{i1} = c_i([s_{i1}]_{\{\text{color,shape}\}})$, (b) choosing the light cube—$s_{i2} = c_i([s_{i2}]_{\{\text{color,shape}\}})$, and (c) choosing the dark ball —$s_{i3} = c_i([s_{i3}]_{\{\text{color,shape}\}})$. Since all pure strategy-indices are symmetric for $v(\{\text{color, shape}\}) = 1$, randomization is suggested by symmetry invariance.

Strategic Game Representations of VUMGs. A VUMG $(M_n, \mathfrak{A}, (\ell_{ia}), v)$ can be represented by a set of strategic forms $(\bar{G}^r)_{r \in \mathbf{R}}$. For all repertoires r, $\bar{G}^r$ is similar to M_n. We set $\bar{I}^r := I$, $\bar{S}_i^r := S_i(r)$, and the payoff functions $\bar{u}_i^r$ as follows: Let $\bar{\theta} \in \bar{\Theta} := \bigtimes_{r \in \mathbf{R}} \Sigma(r)$ be a vector containing mixed-strategy combinations for each $\bar{G}^r$ where $\Sigma(r) := \Sigma_1(r) \times \Sigma_2(r)$. Given $\bar{\theta} \in \bar{\Theta}$, the payoff function in $\bar{G}^r$ is defined by

$$\bar{u}_i^r(\bar{s}) := \sum_{r' \subset r} v(r'|r)u_i(\bar{s}_i, \bar{\theta}_{r',-i}) + v(r|r)u_i(\bar{s}) \tag{2.14}$$

for all $\bar{s} \in S(r)$ and $i \in I$ where $-i$ denotes i's opponent and $\bar{\theta}_{r'-i}$ denotes $-i$'s strategy in $\bar{\theta}$ for $\bar{G}^{r'}$ I.e., $\bar{u}_i^r$ is the expected payoff of player i under the assumption that the subrepertoire players follow $\bar{\theta}$. The first term in (2.14) represents the component of player i's expected payoff concerning the strategies of the subrepertoire players that are given by $\bar{\theta}_{-i}$, i.e., that are fixed in $\bar{G}^r$. In contrast, the second term depends on the choice of the opponent with the same repertoire—r, i.e., player $-i$'s choice for the repertoire r is not determined by $\bar{\theta}_{-i}$ in $\bar{G}^r$.

Based on intuitive assumptions about the players' knowledge and rationality in VUMGs, Bacharach (1993) suggests the following solution of VUMGs. Originally, he omits the uniqueness requirement. In many interesting cases, however, this condition is met, and besides, in case of several undominated equilibria, the players face an equilibrium selection problem which is sidestepped by the stronger version below. Additionally, Bacharach & Stahl (2000) do not employ symmetry invariance as a selection criterion. Instead, they apply some requirement of symmetry disqualification: perceived options the carriers of which have the same cardinality cannot be part of a solution. Clearly, both concepts do not always coincide, but they formalize very similar intuitions. As the coverage functions, both concepts can be viewed as expressions of the principle of insufficient reason.

Definition 2.4.1 (Bacharach solutions of VUMGs).

Consider a VUMG $(M_n, \mathfrak{A}, (\ell_{ia}), v)$ and its strategic game representation based on some coverage function. A vector $\bar{\theta} \in \bar{\Theta}$ is called a Bacharach solution of the VUMG if for all r, $\bar{\theta}_r \in \Sigma(r)$ is the unique symmetry invariant equilibrium of $\bar{G}^r$ that is not dominated by another symmetry invariant equilibrium.

Two different coverage functions have been presented in (2.12) and (2.13). As the following example reveals, they might produce different Bacharach solutions. I.e., Bacharach solutions are very sensitive to the actual specification of the coverage functions.

Example 2.4.3.
Reconsider the game from Example 2.2.1. Assume that both properties—color and shape—come to the mind of the players for sure. So we have one repertoire only—{color, shape}. For the coverage function given by (2.13), the perceived-option set is the same as for the original game. Hence, $\bar{G}^{\{color, shape\}} = M_3$; and the only symmetry invariant strategy-index combination is both players' randomization over all (original) options, which also is the Bacharach solution for this coverage function.

Consider now the coverage function according to (2.12) inducing the five perceived options given in Example 2.4.2. The symmetry invariance requirement of the Bacharach solution then requires the options (b) and (d), and (c) and (e) to be chosen with the same probabilities respectively. Finally, in contrast to the coverage function considered above, the payoff dominance requirement selects randomization over (c) and (e) as the solution.

Janssen's Reduction. Many of the strategic game representations of VUMGs lack Bacharach solutions because of the uniqueness requirement in Definition 2.4.1. This can be seen from the following example.

Example 2.4.4.
Consider a one-shot matching game where two players independently have to choose a physical object from a set consisting of a red ball, a blue ball, two green balls, and three yellow balls.

Assuming that the balls are distinguished by their color only, we have one attribute color with four labels—red, blue, green, and yellow—in the corresponding VUMG. According to the coverage function (2.12), the players have five perceived options: (a) choosing the red ball, (b) choosing the blue ball, (c) picking a green ball, (d) picking a yellow ball, and (e) randomization. Assuming that the players realize the objects' color for sure, we obtain the payoff matrix for both repertoire-{color} players as in Figure 2.8, where the second component of the strategy-indices' subscripts refers to the perceived options above.

	s_{2a}	s_{2b}	s_{2c}	s_{2d}	s_{2e}
s_{1a}	1	0	0	0	$\frac{1}{7}$
s_{1b}	0	1	0	0	$\frac{1}{7}$
s_{1c}	0	0	$\frac{1}{2}$	0	$\frac{1}{7}$
s_{1d}	0	0	0	$\frac{1}{3}$	$\frac{1}{7}$
s_{1e}	$\frac{1}{7}$	$\frac{1}{7}$	$\frac{1}{7}$	$\frac{1}{7}$	$\frac{1}{7}$

Fig. 2.8. Payoff matrix for repertoire {color}

(a)

	s_{2ab}	s_{2c}	s_{2d}	s_{2e}
s_{1ab}	$\frac{1}{2}$	0	0	$\frac{1}{7}$
s_{1c}	0	$\frac{1}{2}$	0	$\frac{1}{7}$
s_{1d}	0	0	$\frac{1}{3}$	$\frac{1}{7}$
s_{1e}	$\frac{1}{7}$	$\frac{1}{7}$	$\frac{1}{7}$	$\frac{1}{7}$

(b)

	s_{2abc}	s_{2d}	s_{2e}
s_{1abc}	$\frac{1}{4}$	0	$\frac{1}{7}$
s_{1d}	0	$\frac{1}{3}$	$\frac{1}{7}$
s_{1e}	$\frac{1}{7}$	$\frac{1}{7}$	$\frac{1}{7}$

Fig. 2.9. Reduced payoff matrices for repertoire {color}

Obviously, the strategy-indices s_{ia} and s_{ib} are symmetric according to Definition 2.3.9 (p. 19). By Corollary 2.3.1, s_{ia} and s_{ib} get the same probabilities in a symmetry invariant strategy-index combination. Hence, there are two undominated symmetry invariant equilibria in this game—choosing s_{ic} and randomizing over s_{ia} and s_{ib}. Therefore, the VUMG has no Bacharach solution.

Since choosing the red cube (s_{ia}) and choosing the blue cube (s_{ib}) are symmetric, instead of viewing them as separate perceived options, the players may perceive them as a single option s_{iab}—picking s_{ia} or s_{ib}. This way, the payoff matrix from Figure 2.8 is reduced to the payoff matrix in Subfigure 2.9 (a). But now, the strategy-indices s_{iab} and s_{ic} are symmetric, and a similar reduction gives the payoff matrix in Subfigure 2.9 (b), where s_{iabc} denotes randomization over s_{iab} and s_{ic} in Subfigure 2.9 (a). In the original game given in Figure 2.8, s_{iabc} is equivalent to choosing both s_{ia} and s_{ib} with probability $\frac{1}{4}$ and to choosing s_{ic} with probability $\frac{1}{2}$, i.e., the red ball, the blue ball, and both green balls are each taken with probability $\frac{1}{4}$. Now, we have a unique undominated symmetry invariant equilibrium—both players' picking a yellow ball.

The reduction mechanism sketched above can be formalized by the following definitions.

Definition 2.4.2 (Reduced strategic games).
The reduced form red(G) *of* $G = (I, (S_i)_{i \in I}, (u_i)_{i \in I}) \in \mathcal{G}$ *is the strategic*

game $\bar{G} = (I, (\bar{S}_i)_{i \in I}, (u_i)_{i \in I}) \in \mathcal{G}$, where, for all $i \in I$, $\bar{S}_i \subseteq \Sigma_i$ is defined as follows: Let $\sim_i$ be the strategy symmetry relation of Definition 2.3.9 restricted to S_i, let $\mathfrak{S}_i$ denote the set of equivalence classes of S_i induced by $\sim_i$, and let $\mathfrak{s}_i(s_i)$ denote the class of $\mathfrak{S}_i$ containing s_i. Now, we set

$$\bar{S}_i := \{ c_i(\mathfrak{s}_i) | \mathfrak{s}_i \in \mathfrak{S}_i \}.$$

I.e., in the reduced forms of strategic games, symmetric strategy-indices of one player are replaced by their centroid (see Definition 2.3.2, p. 15). The other way round, the pure strategy-indices in the reduced form of a strategic game can be identified with (mixed) strategy-indices of the original game in the straightforward way. One player's symmetric strategy-indices of the original game are forced to be taken with the same probabilities. I.e., the players' strategic uncertainty concerning symmetric strategies is resolved in the reduced game (Crawford & Haller 1990). For example, the game in Subfigure 2.9 (a) is the reduced form of the game in Figure 2.8.

This reduction can be applied repeatedly to the resulting strategic games. Since the number of strategy-indices obviously never increases and since the number of strategy-indices is finite, the reduction process eventually becomes stationary after a finite number of steps.

Definition 2.4.3 (Fully reduced strategic games).
The fully reduced form of $G \in \mathcal{G}$ is the strategic game $\mathrm{red}^*(G) \in \mathcal{G}$ such that

1. $\mathrm{red}^*(G) = \mathrm{red}^n(G)$ *for some $n \in \mathbb{N}$ and*
2. $\mathrm{red}^n(G) = \mathrm{red}^{n+1}(G)$.

Note that $\mathrm{red}^*(G)$ does not depend on the choice of n and therefore is always well-defined. For example, the game in Subfigure 2.9 (b) is the fully reduced form of the game in Figure 2.8. Again, the strategy-indices in the fully reduced form of a strategic game can be identified with strategy-indices in the original game.

In order to remedy the non-uniqueness of Bacharach solutions of VUMGs, Janssen (1997) extends the Bacharach solution by a reduction algorithm that can be expressed in terms of the fully reduced strategic games of Definition 2.4.3.

Definition 2.4.4 (Janssen solutions of VUMGs).
Consider some VUMG $(M_n, \mathfrak{A}, (\ell_{ia}), v)$ and its strategic game representation based on the coverage function in (2.13). A vector $\bar{\theta} \in \bar{\Theta}$ is called a Janssen solution of the VUMG if for all r, $\bar{\theta}_r \in \Sigma(r)$ is the unique strategy-index combination of $\mathrm{red}^*(\bar{G}^r)$ that is not dominated by another strategy-index combination.

In contrast to the Bacharach solution, Janssen does not require the single repertoire components of the solution to be equilibria. However, Janssen

(1997) establishes generic existence, equilibrium, and efficiency properties for his solution.

Let $VUMG(n, \mathfrak{A}, (\ell_{ia}))$ be the class of all VUMGs $(M_n, \mathfrak{A}, (\ell_{ia}), v)$ based on fixed n, $\mathfrak{A}$, and (ℓ_{ia}). Each VUMG of this class is characterized by the vector v of the repertoire probabilities $v(r)$, and the whole class can be represented by a $|\mathbf{R}| - 1$-dimensional simplex $\mathcal{V}$ of the $|\mathbf{R}|$-dimensional Euclidean space. A property for this class is said to hold generically or in generic cases if there is a null set $\mathcal{N} \subset \mathcal{V}$ with respect to the $|\mathbf{R}| - 1$-dimensional Lebesgue measure such that the property holds for all VUMGs corresponding to $\mathcal{V} \backslash \mathcal{N}$.

Theorem 2.4.1 (Janssen 1997).
Let $(M_n, \mathfrak{A}, (\ell_{ia}), v) \in VUMG(n, \mathfrak{A}, (\ell_{ia}))$ be a VUMG and let $(\bar{G}^r)_{r \in \mathbf{R}}$ its strategic game representation based on the coverage function in (2.13) according to Subsubsection 2.4.2.

1. *In generic cases, $(M_n, \mathfrak{A}, (\ell_{ia}), v)$ has a Janssen solution $\theta^* \in \Theta$.*
2. *If θ^* is a Janssen solution of $(M_n, \mathfrak{A}, (\ell_{ia}), v)$ then θ_r^* is an equilibrium of $\bar{G}^r$ for all $r \in \mathbf{R}$.*
3. *In generic cases, a focal point θ^* of $(M_n, \mathfrak{A}, (\ell_{ia}), v)$ has the following property: If $\mathfrak{a} \in r$ and $\bar{u}_i^{\{\mathfrak{a}\}}(\theta_{\{\mathfrak{a}\}}^*) > \frac{1}{n}$ then $\bar{u}_i^r(\theta_r^*) > \frac{1}{n}$ for all $r \in \mathbf{R}$ and $\mathfrak{a} \in \mathfrak{A}$.*

In view of Definition 2.4.4, the first part secures the generic uniqueness of Janssen solutions—in almost all cases, (unique) Janssen solutions exist in VUMGs. The second part shows that Janssen solutions are stable in the sense that no player has an incentive to deviate from it, not even by ignoring the symmetry of the pure strategies and choosing a mixed strategy that is not symmetric. The third part states a generic efficiency property of Janssen solutions: If players with a simple repertoire consisting of just one attribute can do better than randomization—payoff $\frac{1}{n}$—then all players with repertoires that contain this attribute can also reach a higher payoff than by randomizing. As Janssen (1997) points out, this theorem explains the phenomenon that players coordinate much more frequently in matching games than by randomizing. Of course, if two players with different repertoires are matched, they might fail to coordinate.

Experimental Results. Bacharach & Bernasconi (1997) test the variable universe approach in controlled experiments for some matching games. A major problem in doing this concerns the distribution of the repertoires: Given a certain coverage function, many outcomes can be supported by assuming different distributions of repertoires. I.e., it is difficult to falsify the variable universe approach. Bacharach & Bernasconi solve this problem by trying to find out the distribution of repertoires independently of the players' observed choices. For this purpose, they exploit the results of earlier experiments with the same players as well as the players' written reports.

	s_{2a}	s_{2b}	s_{2c}	s_{2d}	s_{2e}
s_{1a}	$\frac{1}{20}$	$\frac{1}{20}$	$\frac{1}{20}$	$\frac{1}{20}$	$\frac{1}{20}$
s_{1b}	$(1-\bar{v})\frac{1}{2}+\bar{v}\frac{1}{20}$	$\frac{1}{2}$	$(1-\bar{v})\frac{1}{2}$	$(1-\bar{v})\frac{1}{2}$	$(1-\bar{v})\frac{1}{2}+\bar{v}\frac{1}{19}$
s_{1c}	$\bar{v}\frac{1}{20}$	0	$\bar{v}\frac{1}{18}$	$\bar{v}\frac{1}{18}$	$\bar{v}\frac{17}{18\cdot19}$
s_{1d}	$\bar{v}\frac{1}{20}$	0	$\bar{v}\frac{1}{18}$	$\bar{v}$	0
s_{1e}	$(1-\bar{v})\frac{1}{19}+\bar{v}\frac{1}{20}$	$\frac{1}{19}$	$(1-\bar{v})\frac{1}{19}+\bar{v}\frac{17}{18\cdot19}$	$(1-\bar{v})\frac{1}{19}$	$\frac{1}{19}$

Fig. 2.10. Payoff matrix for $\bar{G}^{\{\texttt{color,grain}\}}$

Based on this empirical distribution, the experimental results support the variable frame approach to a large extent. More precisely, Bacharach & Bernasconi establish a joint set of hypotheses, among them, besides the above mentioned derivation of distribution of repertoires, also the way of how the players perceive their options (e.g., the realized attributes), and, of course, their interaction within the variable frame theory.

An Example. Reconsider the game from Example 2.2.2 (p. 12). This game can be modelled as a strategic game representation of a VUMG based on the matching game M_{20}. Without loss of generality, the labelling is given by $\ell_{i\texttt{color}}(s_{ik}) = \texttt{red}$ for $k \in \{1,2\}$ and $\ell_{i\texttt{color}}(s_{ik}) = \texttt{yellow}$ for $k \in \{3,\dots,20\}$ and $i \in I$; and $\ell_{i\texttt{grain}}(s_{i3}) = \texttt{wavy}$ and $\ell_{i\texttt{grain}}(s_{ik}) = \texttt{straight}$ for $k \in \{1,2,4,\dots,20\}$ and $i \in I$. We consider two repertoires only—$\{\texttt{color}\}$ and $\{\texttt{color,grain}\}$, i.e., we have $v(\{\texttt{color}\}) = 1 - \bar{v}$ and $v(\{\texttt{color,grain}\}) = \bar{v}$ for some $\bar{v} \in [0,1]$. Furthermore, we assume coverage functions according to (2.12).

In $\bar{G}^{\{\texttt{color}\}}$, the perceived-option set $\bar{S}_i^{\{\texttt{color}\}}$ comprises three strategies— (a) randomization over all blocks, (b) randomization over all red blocks, and (c) randomization over all yellow blocks. Obviously, randomizing over the red blocks is the $\bar{G}^{\{\texttt{color}\}}$-part in the Bacharach solution.

In $\bar{G}^{\{\texttt{color,grain}\}}$, the perceived-option set $\bar{S}_i^{\{\texttt{color,grain}\}}$ comprises two more strategies—(d) take the (yellow) block with the wavy grain of wood and (e) randomization over all blocks with a straight grain of wood. In view of the $\bar{G}^{\{\texttt{color}\}}$-part of the Bacharach solution, we have the payoff matrix for player 1 in $\bar{G}^{\{\texttt{color,grain}\}}$ as in Figure 2.10, where a player's single strategies are indexed as above; player 2's payoff matrix is the transpose of this matrix. As indicated in Section 2.2, for $\bar{v} > \frac{1}{2}$, taking the wavy yellow block, and for $\bar{v} < \frac{1}{2}$, picking a red block is the $\bar{G}^{\{\texttt{color,grain}\}}$-part of the Bacharach solution.

2.4.3 Sugden's Labelling

Bacharach (1991, 1993), Janssen (1997), Bacharach & Bernasconi (1997), and Bacharach & Stahl (2000) invoke a *principle of insufficient reason* which has the effect of ruling out as a solution any strategy which is symmetric—considering the description—to another strategy. In contrast, Sugden (1995) claims to do without. In his approach, the players are somehow or other given a unique one-dimensional private label for each strategy. In a shared culture, the assignment of labels is correlated across players. Sugden explores ways in which players can exploit this correlation to increase the chance of coordination in coordination problems, for example, by following a rule which selects the same label for all of them.

In addition, Sugden admits uncertainty concerning the labelling. Since the players identify the strategies by their labels only, this uncertainty may restrict the set of strategies (strategy combinations) that players consider to be implementable. This aspect of Sugden's labelling is related to Crawford & Haller (1990, p. 580) who assume that the "[p]layers' descriptions of the game are privately observed permutations of the analyst's labellings of players' positions and actions ... (that) are drawn from commonly known distributions, which assign equal prior probability to alternative labellings of undistinguished positions or actions, but always assign the same label for each player to distinguished positions or actions ...".

Example 2.4.5 (Sugden 1995).
Two players are each given a bag containing five discs. The discs are all identical except for that three of them are red and two of them are blue. In addition, the discs are numbered, but such that the numbers can realized by the game master only. Suppose that the red discs are numbered by 1, 2, and 3, and the blue ones are numbered by 4 and 5. Independently, the players have to take the discs out of the bag, and then have to chose one of them. If the players have chosen the same disc, both get a prize, otherwise they get nothing.

Suppose both players distinguish the discs in terms of two features: their color and equally colored discs by their order in which they were drawn from the bag. This way, all of one player's discs get different labels. However, there is some degree of uncertainty concerning this labelling. For example, the red disc 1 can get one of the labels red&first, red&second, or red&third—depending on the order of drawing. Moreover, the players may label the same disc differently—player 1 may label the red disc 1 by red&first, while player 2 labels this disc by red&second.

Since the players can distinguish the strategies by their private description only, the setting of the game restricts the strategies the players think to be implementable. Supposing that all pairs of drawing orders are equally likely, choosing for example the disc labelled red&first induces the mixed strategy of randomizing over all red discs. Of course, the same holds for taking the disc labelled red&second. Additionally, the players cannot, say, coordinate by each taking the disc labelled red&first as this label refers to different discs with probability $\frac{2}{3}$. Thus, in effect, the players have two pure strategies only: picking a red disc or picking a blue disc. The principle of coordination then selects both players' picking any of the blue discs as a solution.

Viewing the drawing procedure as a real one, Sugden merely considers games with modified strategy spaces. In our example above, he does *not* consider the original matching games M_n. Actually, he considers similar coordination games with reduced strategy spaces—to pick a red disc or to pick a blue disc instead of choosing any of the discs in the example above. Since by this setting the players are *forced* to randomize over strategies with a similar labelling, i.e., the color red, he can do without the principle of insufficient reason. So, as in Grant & Quiggin (1998), the main message is that we must be careful about the actual strategy spaces.

Another interpretation of the drawing procedure seems to be conceivable: Suppose all discs lie on a table from the beginning of the game. Then, the discs are not drawn in reality, but they may be drawn in the players' minds. It seems to be quite plausible to assume that all pairs of these 'mental drawing' orders are equally probable. Of course, this assumption is nothing less than some expression of the principle of insufficient reason—options with a similar description are to be taken with the same probabilities. Hence, Sugden's labelling can be viewed as a way of justifying the coverage functions from Subsection 2.4.2 (pp. 27).

2.4.4 The Principle of Coordination

All formalizations of focal points/salience invoke one or another version of the principle of coordination. In its strong version, it involves a uniqueness requirement—the 'solution' is the unique undominated (Pareto-efficient) equilibrium of the game under consideration. Gauthier (1975), Sugden (1995), Janssen (1997), and Bacharach (1998) employ this strong version. To be precise, Janssen (1997) applies a *principle of individual team member rationality* as an individualistic variant of the principle of coordination—the players do their part in such a solution. But this way, a solution may not exist. In contrast, Bacharach (1993) and Bacharach & Bernasconi (1997) do without uniqueness. Their weak version requires a solution to be undominated only. While always ensuring a solution, quite often, this version leaves the players with a multitude of recommended solutions; therefore causing a serious selection problem.

The principle of coordination is not uncontroversial. Gilbert (1990), Sugden (1991), and Hollis & Sugden (1993) object that the principle of coordination cannot be supported by common knowledge of the structure of the game and players' rationality alone: Consider the coordination game in Figure 2.11. Obviously, (s_{11}, s_{21}) dominates the other equilibria. However, taking the dominated equilibrium (s_{12}, s_{22}) does not contradict common knowledge of the structure of the game and of the players' rationality—given the players' expectations of the other player's choice by (s_{12}, s_{22}), s_{12} and s_{22} are

	s_{21}	s_{22}
s_{11}	2 2	0 0
s_{12}	0 0	1 1

Fig. 2.11. A coordination game

their rational choices respectively. More generally, all rationalizable strategies (Bernheim 1984; Pearce 1984) are consistent with common knowledge of rationality.

Gilbert (1990) points out that a payoff dominant equilibrium might be risk dominated (see Harsanyi & Selten 1988); and therefore, it is not clear whether it should be taken or not. However, Harsanyi & Selten (1988) give payoff dominance priority over risk dominance within their general solution concept.

Schlag (1993) and Anderlini (1999) provide some justification for the principle of coordination for common interest games that are extended by cheap talk. In order to do this, Schlag employs evolutionary game theory. In contrast, Anderlini imposes computability constraints on the players' strategies and the occurring trembles.

Another way to support the principle of coordination is suggested by Sugden (1993): Under certain circumstances, for example in coordination games, players may conceive themselves as members of team, and therefore do not reason individually. Then, the question to be answered is: Which rule, if followed by both, would give the best result for both? In common interest games, obviously, to do one's part in unique payoff dominant equilibrium is such a rule. In view of the above arguments, "[a]ll what we can say, then, is that the Principle of Coordination is a principle of rationality for players who conceive of themselves as a team, but not for players who do not" (Hollis & Sugden 1993, p. 13).

Insofar the arguments in favor of the choice of a salient option hinge upon the payoff dominance principle, all objections against this principle also are valid against these arguments. Gilbert's (1989) criticism is directed against reconceiving a game based on the salience of a certain option as suggested by Gauthier (1975). In the coordination games under consideration, the mutual expectation of choosing some (distinguished) non-salient option is consistent and, of course, leads to the choice of that option. In order to make this argument work, it is not even necessary to assume some kind of salience-shyness (Gilbert 1989), i.e., some aversion of a player to take a salient option. Note that in order to formalize this aversion in a game, the player's preferences had to be redefined.

2.5 Framed Strategic Games (FSGs)

In order to formalize focal point arguments, this section extends strategic games by a structure called *frame* (Subsection 2.5.1). This structure adds information about the players' descriptions of the game via a multidimensional labelling of the strategies. Since there is some degree of arbitrariness in making up these frames, an invariance requirement on the solution of these 'framed' games, similar to invariance with respect to isomorphisms in strategic games,—*language invariance*—is introduced (Subsection 2.5.2). Though being quite strong, this language invariance requirement often breaks the symmetry of strategy-indices in the original game and singles out more efficient solutions—focal points. Subsection 2.5.3 derives symmetries and symmetry invariance from isomorphisms and language invariance and explores some of their properties. Limitations of the FSG approach are discussed in Subsection 2.5.4. Subsection 2.5.5 deals with a peculiar feature of frames. Some applications and implications of the FSG approach are given in the final subsection.

2.5.1 Frames

The players' description of strategies in strategic games is formalized by frames as follows.

Definition 2.5.1 (Frames of strategic games).
A frame F of a strategic game $G = (I, (S_i)_{i \in I}, (u_i)_{i \in I})$ is a triple $(\mathfrak{A}, \Lambda, (\ell_{i\mathfrak{a}})_{(i,\mathfrak{a}) \in I \times \mathfrak{A}})$, where $\mathfrak{A}$ denotes a non-empty set of attributes $\mathfrak{a}$, Λ a non-empty set of labels λ, and $\ell_{i\mathfrak{a}}$ the label function $\ell_{i\mathfrak{a}} : S_i \to \Lambda$ for each player $i \in I$ and attribute $\mathfrak{a} \in \mathfrak{A}$.

The pair $(\mathfrak{A}, \Lambda)$ is called the language of a frame F and $(\ell_{i\mathfrak{a}})_{(i,\mathfrak{a}) \in I \times \mathfrak{A}}$ its labelling. While the language is independent of the game, the labelling connects the game and the language—each strategy-index s_i is assigned a multidimensional label $(\ell_{i\mathfrak{a}}(s_i))_{\mathfrak{a} \in \mathfrak{A}}$. The attributes $\mathfrak{a}$ (e.g., color or shape) can be thought of as representing properties of the strategies (e.g., color or shape of an object to choose) as they are perceived by the players, and the labels λ (e.g., red or cubic) characterize the individual strategies with respect to one property (e.g., the redness of an object). A frame $F = (\mathfrak{A}, \Lambda, (\ell_{i\mathfrak{a}})_{(i,\mathfrak{a}) \in I \times \mathfrak{A}})$ is called finite if both $\mathfrak{A}$ and Λ are finite.

Example 2.5.1.
The perception of Example 2.2.1 (p. 10) by players who are aware of color and shape but nothing else can be modelled by the strategic game M_3 and the frame F where $\mathfrak{A} = \{\text{color}, \text{shape}\}$, $\Lambda = \{\text{light}, \text{dark}, \text{round}, \text{cubic}\}$, and $(\ell_{i\mathfrak{a}})_{(i,\mathfrak{a}) \in I \times \mathfrak{A}}$ as given by Figure 2.12 for all $i \in I$.

s_{ij} $a \quad \ell_{ia}(s_{ij})$	s_{i1}	s_{i2}	s_{i3}
color	light	light	dark
shape	round	cubic	round

Fig. 2.12. A frame of a strategic game

Note, it is the *players'* perception of a game that is represented by a frame. It is taken for granted that game theorists know the players' perception of the game, and that their (formal) language $(\mathfrak{A}, \Lambda)$ is rich enough to cover this perception. Swahili, for example, might contain more words for expressing shades of green than English. In this case, an English-speaking game theorist possibly will use some Swahili words in order to formalize the perception of a game by a Swahili-speaking player.

The term *frame* indicates some relation to Tversky & Kahneman's (1981) and Kahneman & Tversky's (1984) frames and framing effects in decision theoretic contexts. They observe that different framings of acts lead to different choices even if the decision theoretic structure does not change. Whereas decision makers (players against nature) respond to the framing of a decision problem rather unconsciously, the players in a game may consciously exploit a frame to their favor. Nevertheless, focal points in games can be viewed as an expression of framing effects in games. For example, the pure coordination problems in Figure 1.2 (p. 3) and Figure 1.3 (p. 4) can be viewed as different framings of the strategic game M_3 (see Figure 1.1, p. 2)—both games differ in the players' descriptions of their actions only. As indicated in Subsection 1.1.1, focal point reasoning might lead the players to choose different actions in the two situations. Rottenstreich (1996) considers framing effects in games that do not rest upon breaking symmetries. In his experiments, he observes for example that the labelling of players matters in bargaining situations, even if their positions are strategically equivalent. Bacharach & Stahl (2000) use the term frame to denote a players possibilities to perceive a game, i.e. the set of properties of the strategies that a player actually realizes—some subset of the attribute set $\mathfrak{A}$.

The formalization of the players' apprehension of a game by frames is closely related to Bacharach's (1993) families of concepts—the attributes being associated with the families and the labels with the concepts. While concepts belong to one family exclusively, in principle, one label could be assigned with respect to several attributes. Formally, we allow a label to be assigned with regard to several attributes. This wider setting proves useful when, for example, a frame describes the color of different parts of an object to choose (see Subsection 2.5.5). A further difference to Bacharach (1991) is that the labels of a frame are exhausting—each strategy is actually assigned a label for each attribute. The gap to non-exhausting families can easily be filled by adding the label no_label_assigned to the label set and assigning it to the strategies that are not covered by a concept. The possibility that a player does not realize certain families, admitted in VFT, can be accounted for by a label `attribute_not_realized_by_this_player` that is assigned to all of this player's strategy-indices. This way all strategy-indices of this player remain undistinguished with respect to this attribute. Another basic feature of VFT is the players' uncertainty concerning the other players' apprehension of a game. It can be modelled by FSGs where the player-indices are the type agents of a Bayesian game in which a player's type agents will differ both in their labelling of the strategy-indices and in their expectations over the other players' types (labellings). While Sugden's (1995) label functions assign a unique one-dimensional denotation to a player's strategies, in FSGs, each strategy obtains a multidimensional label that is not necessarily unique. The uncertainty with respect to both the own and the other players' label functions can be handled in the same way as for the VFT.

Definition 2.5.2 (Framed strategic games).
A pair (G, F) is called framed strategic game (FSG). A FSG (G, F) is finite if both G and F are finite; $\mathcal{F}$ denotes the class of finite FSGs.

Equilibria of FSGs are the equilibria of the underlying strategic games. The following extension of solution concepts for strategic games makes explicit the idea that different framings of a given strategic game may result in different solutions:

Definition 2.5.3 (Solution concepts for FSGs).
A solution concept L for $\mathcal{F}$ assigns a set of strategy-index combinations $L(G, F) \subseteq \Sigma$ to every FSG $(G, F) \in \mathcal{F}$.

As all other rationalizations of focal points mentioned above, the FSG approach concentrates on the players' labelling of strategies; properties of players and payoffs (results, prizes) are not dealt with directly. Nevertheless, it is possible to label the players themselves. This can be done by introducing attributes that characterize players, like `sex_of_player` or `color_of_player` and adding appropriate labels and label functions. Since player attributes should not distinguish a player's strategy-indices, all strategy-indices of one player are to be assigned the same label with respect to player attributes. An application of player labels is given below (pp. 49).

As for payoff attributes, one could argue that there is no need for them: For example, a single worst paying object in a matching game is sufficiently distinguished by the payoff structure. If all other objects give the same payoff in the case of coordination, then they are symmetric and should be assigned the same probabilities. This way it may be possible that payoff dominance selects the single worst-paying object as the solution, even without taking into account the players' labelling. On the other hand, we could, of course, introduce payoff attributes and assign the appropriate labels to the strategy-indices that lead to the respective payoffs.

	s_{ij}	s_{i1}	s_{i2}	s_{i3}
$\mathfrak{a}$ $\bar{\ell}_{i\bar{\mathfrak{a}}}(s_{ij})$				
Farbe		hell	hell	dunkel
Form		rund	eckig	rund

Fig. 2.13. Another frame of a strategic game

2.5.2 Language Invariance

Isomorphisms of FSGs. In general, two game theorists—especially game theorists with different native languages—who formalize the same decision problem will use different expressions to denote the players' attributes and labels. While Figure 2.12 shows the frame of an English-speaking game theorist for Example 2.2.1, a German-speaking game theorist will probably set up the frame $\bar{F}$ for M_3 where $\bar{\mathfrak{A}} = \{\text{Farbe,Form}\}$, $\bar{\Lambda} = \{\text{hell,dunkel,rund, eckig}\}$, and $(\bar{\ell}_{i\mathfrak{a}})_{(i,\bar{\mathfrak{a}})\in\bar{I}\times\bar{\mathfrak{A}}}$ is given by Figure 2.13 for all $i \in I$.

Since the resulting FSGs represent the same decision problem, they are similar in a very precise sense: (a) The underlying strategic games are isomorphic. (b) The labellings can be transformed into each other by translating one game theorist's (formal) language into the other game theorist's (formal) language. In the example above, this can simply be done by translating the attributes and labels of F which are words of the natural language English into German. For example, the attribute color from F is translated into the attribute Farbe in F'. This similarity is formalized with the following definition:

Definition 2.5.4 (Isomorphisms of FSGs).
Let $(G, F), (\bar{G}, \bar{F}) \in \mathcal{F}$ be FSGs $((I, (S_i)_{i\in I}, (u_i)_{i\in I}), (\mathfrak{A}, \Lambda, (\ell_{i\mathfrak{a}})_{(i,\mathfrak{a})\in I\times\mathfrak{A}}))$ and $((\bar{I}, (\bar{S}_i)_{i\in\bar{I}}, (\bar{u}_i)_{i\in\bar{I}}), (\bar{\mathfrak{A}}, \bar{\Lambda}, (\bar{\ell}_{i\bar{\mathfrak{a}}})_{(i,\bar{\mathfrak{a}})\in\bar{I}\times\bar{\mathfrak{A}}}))$. An isomorphism from (G, F) to $(\bar{G}, \bar{F})$ is an isomorphism $\mathbf{f} = (\pi, (r_i)_{i\in I})$ from G to $\bar{G}$ together with a pair of bijective mappings $\mathbf{t} = (\mu, \tau)$, $\mu : \mathfrak{A} \to \bar{\mathfrak{A}}$ and $\tau : \Lambda \to \bar{\Lambda}$, such that

$$\tau(\ell_{i\mathfrak{a}}(s_i)) = \bar{\ell}_{\pi(i)\mu(\mathfrak{a})}(r_i(s_i)) \tag{2.15}$$

holds for all $i \in I$, $s_i \in S_i$, and $\mathfrak{a} \in \mathfrak{A}$.

Obviously, a system of identity mappings is an isomorphism from a FSG to itself. Since all mappings involved are bijective, inverses and compositions of isomorphisms are also isomorphisms. Thus, isomorphisms constitute an equivalence relation on $\mathcal{F}$. Two FSGs are called isomorphic if they can be mapped onto each other by an isomorphism.

Example 2.5.2.
Consider the FSF representations of Example 2.2.1 given by the FSG (M_3, F) in Example 2.5.1 and the FSG $(M_3, \bar{F})$ as defined above. Then, $(\mathbf{f}, \mathbf{t}) =$

$\mathfrak{a}$	s_{ij}	$\ell_{i\mathfrak{a}}(s_{ij})$	$\tau(\ell_{i\mathfrak{a}}(s_{ij}))$		$\ell_{\pi(i)\mu(\mathfrak{a})}(r_i(s_{ij}))$	$\mu(\mathfrak{a})$	$r_i(s_{ij})$
color	s_{i1}	light	hell	=	hell	Farbe	s_{i1}
color	s_{i2}	light	hell	=	hell	Farbe	s_{i2}
color	s_{i3}	dark	dunkel	=	dunkel	Farbe	s_{i3}
shape	s_{i1}	round	rund	=	rund	Form	s_{i1}
shape	s_{i2}	cubic	eckig	=	eckig	Form	s_{i2}
shape	s_{i3}	round	rund	=	rund	Form	s_{i3}

Fig. 2.14. Checking an isomorphism of FSGs

$((\pi, (r_i)_{i \in I}), (\mu, \tau))$ with $\pi = \mathrm{id}$, $r_1 = \mathrm{id}$, $r_2 = \mathrm{id}$, $\mu(\text{color}) = \text{Farbe}$, $\mu(\text{shape}) = \text{Form}$, $\tau(\text{light}) = \text{hell}$, $\tau(\text{dark}) = \text{dunkel}$, $\tau(\text{round}) = \text{rund}$, and $\tau(\text{cubic}) = \text{eckig}$ constitutes an isomorphism from (M_3, F) to $(M_3, \bar{F})$. This can be seem from Figure 2.14 where $i = 1, 2$.

Definition 2.5.4 is an extension of Definition 2.3.6 to FSGs: Two FSGs are isomorphic if the underlying strategic games are isomorphic and if the labellings can be transformed into each other by translation of the language where this transformation can be done by separate translations of the attributes (μ) and the labels (τ) such that the relation between the payoff structure and the players' label structure is preserved (Condition (2.15)): The translated label of strategy s_i with respect to attribute $\mathfrak{a}$ (left side) has to be equal to the label of the renamed strategy $r_i(s_i)$ with respect to the translated attribute $\mu(\mathfrak{a})$ in the other game (right side).

Invariance with Respect to Isomorphisms. Arguments as those for the solutions of strategic games (see Subsection 2.3.4), support the requirement that isomorphic FSGs should be assigned the same solution. I.e., the solutions of FSGs should not depend on the language $(\mathfrak{A}, \Lambda)$—the translation of the players' strategy descriptions into another (game theorist's formal) language should leave the solution unaffected. This way the solution for a decision problem would be always independent of the game theorist's description. An invariance requirement similar to Definition 2.3.7 formalizes this idea.

Definition 2.5.5 (Language invariance).
A solution concept L for $\mathcal{F}$ is invariant with respect to isomorphisms (language invariant) if for every isomorphism $(\mathbf{f}, \mathbf{t})$ from $(G, F) \in \mathcal{F}$ to $(\bar{G}, \bar{F}) \in \mathcal{F}$ we have $f(L(G, F)) = L(\bar{G}, \bar{F})$ for f given by $\mathbf{f}$ via (2.6).

Language invariance involves the translation of the game theorists' (formal) languages $(\mathfrak{A}, \Lambda)$ as one part of the isomorphisms it rests upon. While natural languages are not always smoothly translatable into each other, the language parts of frames are very simple ones, and there should be no problem in translating them. 'Difficulties' may arise from the fact that the label sets

a	s_{ij}	$\ell_{ia}(s_{ij})$	$\tau(\ell_{ia}(s_{ij}))$		$\ell_{\pi(i)\mu(a)}(r_i(s_{ij}))$	$\mu(a)$	$r_i(s_{ij})$
color	s_{i1}	light	round	=	round	shape	$s_{\pi(i)1}$
color	s_{i2}	light	round	=	round	shape	$s_{\pi(i)3}$
color	s_{i3}	dark	cubic	=	cubic	shape	$s_{\pi(i)2}$
shape	s_{i1}	round	light	=	light	color	$s_{\pi(i)1}$
shape	s_{i2}	cubic	dark	=	dark	color	$s_{\pi(i)3}$
shape	s_{i3}	round	light	=	light	color	$s_{\pi(i)2}$

Fig. 2.15. Checking a symmetry of a FSG

of isomorphic FSGs have to be of the same cardinality, although the range of actually used labels is smaller. Since it is easy to check that isomorphic FSGs have the same number of used labels, one could avoid these difficulties by considering non-redundant FSGs only, i.e., FSGs without non-assigned labels.

2.5.3 Symmetry Invariance

Symmetries. As in strategic games, isomorphisms from a FSG to itself are called symmetries. Next, an example of a symmetry of a FSG is given. Consider Example 2.5.1. Obviously, $\mathbf{f} = (\pi, (r_i)_{i \in I})$ given by any π and $r_i(s_{i1}) = s_{\pi(i)1}$, $r_i(s_{i2}) = s_{\pi(i)3}$, and $r_i(s_{i3}) = s_{\pi(i)2}$ for all i is a symmetry of M_3. Together with $\mathbf{t} = (\mu, \tau)$ given by $\mu(\text{color}) = \text{shape}$, $\mu(\text{shape}) = \text{color}$, $\tau(\text{light}) = \text{round}$, $\tau(\text{dark}) = \text{cubic}$, $\tau(\text{round}) = \text{light}$, $\tau(\text{cubic}) = \text{dark}$ $\mathbf{f}$ constitutes a symmetry $(\mathbf{f}, \mathbf{t})$ of (M_3, F). This is immediate from Figure 2.15 where $i = 1, 2$.

Less formally, this can be justified as follows: English-speaking game theorists might index and label the game in accordance with Figure 2.12. For concreteness, let the player indexed 1 be Bob and the player indexed 2 be Max. So, for example, Max' strategy of taking the light cube obtains the strategy-index s_{22} and is labeled (in English) $(\text{light}, \text{cubic})$. Now imagine a language—Glishen—in which shape means color, color means shape, cubic means dark and vice versa, round means light and vice versa. Imagine a Glishen-speaking game theorist T who indexes Bob 2 and Max 1; and who indexes the light ball strategies s_{i1}, but the light cube strategies s_{i3} and the dark ball strategies s_{i2} for both i. Then T's actual coding (indexing plus labelling) will be the same as ours. But for example, the strategy indexed by T as s_{12} (Max' dark ball strategy) is labeled in Glishen $(\text{light}, \text{cubic})$. This shows that re-indexing plus translation into Glishen make up a symmetry of the game.

Invariance. Symmetry invariant strategy-index combinations in FSGs are defined in analogy to their counterparts in strategic games (see Definition

2.3.8, p. 19). Again, symmetry invariance of a solution within a one-point solution concept is a direct consequence of invariance with respect to isomorphisms.

Definition 2.5.6 (Symmetry invariance in FSGs).
A strategy-index combination $\sigma \in \Sigma$ of a FSG (G, F) is symmetry invariant if $f(\sigma) = \sigma$ for all mappings f given by the strategy parts $\mathbf{f}$ of symmetries $(\mathbf{f}, \mathbf{t})$ of (G, F) via (2.6).

Symmetric Strategies, Attributes, and Labels. The notion of symmetric strategies can be adapted to FSGs in a natural way. Again, symmetric strategies are assigned the same probabilities in all symmetry invariant strategy combinations. In a similar way, the symmetry of attributes and labels is defined.

Definition 2.5.7 (Symmetric strategies, attributes, and labels).
Let $(G, F) = ((I, (S_i)_{i \in I}, (u_i)_{i \in I}), (\mathfrak{A}, \Lambda, (\ell_{i\mathfrak{a}})_{(i,\mathfrak{a}) \in I \times \mathfrak{A}})) \in \mathcal{F}$, $s_i \in S_i$, $s_{i'} \in S_{i'}$, $\mathfrak{a}, \mathfrak{a}' \in \mathfrak{A}$, $\lambda, \lambda' \in \Lambda$, and let $(\mathbf{f}, \mathbf{t}) = ((\pi, (r_i)_{i \in I}), (\mu, \tau))$ be an arbitrary symmetry of (G, F). Then, (a) the strategies s_i and $s_{i'}$, (b) the attributes $\mathfrak{a}$ and $\mathfrak{a}'$, (c) the labels λ and λ' are called symmetric if there is a symmetry $(\mathbf{f}, \mathbf{t})$ of (G, F) such that (a) $r_i(s_i) = s_{i'}$, (b) $\mu(\mathfrak{a}) = \mathfrak{a}'$, (c) $\tau(\lambda) = \lambda'$, respectively.

The example on pp. 42 raises the question what structural properties of frames allow two strategy-indices to be symmetric. The corollary below gives a result which—giving a necessary condition—goes some way to answering this question. In addition, it establishes some more properties of the symmetry relations in FSGs that prove useful in the determination of symmetry invariant strategy-index combinations.

Corollary 2.5.1.
In FSGs, (a) symmetric strategies and (b) symmetric attributes are assigned the same number of different labels, and (c) symmetric labels are assigned to the same number of strategy-attribute pairs.

Proof. Suppose $s_{i'} \in S_{i'}$ and $s_{i''} \in S_{i''}$ from (G, F) are symmetric. By Definition 2.5.7, there is a symmetry $((\pi, (r_i)_{i \in I}), (\mu, \tau))$ of (G, F) such that $r_{i'}(s_{i'}) = s_{i''}$ and $\pi(i') = i''$. For all $i \in I$ and $s_i \in S_i$, let $\Lambda(s_i)$ denote s_i's label set, $\Lambda(s_i) := \{\lambda | \exists \mathfrak{a} \in \mathfrak{A} : \ell_{i\mathfrak{a}}(s_i) = \lambda\}$. For any label $\lambda \in \Lambda(s_{i'})$, there is an attribute $\mathfrak{a}$ such that $\ell_{i'\mathfrak{a}}(s_{i'}) = \lambda$. By Definition 2.5.4 we have $\tau(\lambda) = \tau(\ell_{i'\mathfrak{a}}(s_{i'})) = \ell_{\pi(i')\mu(\mathfrak{a})}(r_{i'}(s_{i'})) = \ell_{i''\mu(\mathfrak{a})}(s_{i''}) \in \Lambda(s_{i''})$. Therefore, $\tau(\Lambda(s_{i'})) \subseteq \Lambda(s_{i''})$. Since τ is bijective we have $|\Lambda(s_{i'})| \leq |\Lambda(s_{i''})|$. Since inverses of symmetries are symmetries too, one can show that $|\Lambda(s_{i'})| \geq |\Lambda(s_{i''})|$ in a similar way. Thus, we have $|\Lambda(s_{i'})| = |\Lambda(s_{i''})|$.

The proof of the next two claims follows very much the idea of the proof of the first one and is therefore omitted. $\square$

Reconsider Example 2.2.1. By the symmetry shown in Subsection 2.5.3 and by letting π be the identity mapping, it follows that the second and the third strategy of both players are symmetric. It is also quite obvious—having $\pi(1) = 2$ and $\pi(2) = 1$—that s_{1j} and s_{2j} are symmetric for $j = 1, 2, 3$. But s_{11} and s_{12} cannot be symmetric: According to Definition 2.5.4, a symmetry $(\mathbf{f}, \mathbf{t})$ that constitutes the symmetry of s_{11} and s_{12} would have to map either the label light or the label round to the label cubic or vice versa. Since in contrast to cubic, which is assigned to two strategy-attribute pairs only, both light and round are assigned to four strategy-attribute pairs, by Corollary 2.5.1, this is impossible. Thus in a symmetry invariant strategy combination, both players take the 'same' strategy and assign the same probability to the second and the third strategy.

Existence of Symmetry Invariant Equilibria. By definition, the strategy part $\mathbf{f}$ of a symmetry $(\mathbf{f}, \mathbf{t})$ of a FSG (G, F) is a symmetry of the underlying strategic game G. Since Definition 2.5.4 imposes additional restrictions, symmetry invariance in FSGs involves a smaller set—at most the same set—of symmetries $\mathbf{f}$ to be considered than symmetry invariance in the underlying strategic games does. Generally, symmetry invariance in FSGs is weaker than in strategic games. Thus, Theorem 2.3.3 implies the existence of symmetry invariant equilibria for finite FSGs.

Theorem 2.5.1.
Any finite FSG has a symmetry invariant equilibrium.

2.5.4 Meaning and Structure of Labels—an Extraterrestrial's Perspective

A limitation of the FSGs approach is that language invariance ignores the inner structure of the labels and the meaning that players may attach to strategy descriptions. This is explained with some examples:

Consider a matching game with three differently colored but otherwise identical cubes—(1) a green one, (2) a brown one, and (3) an ochre one. The game can be represented by a FSG with the single attribute color and the label set {green, brown, ochre}. Supposing that there is nothing special about any of these colors, randomization is the intuitive solution. This is also the only symmetry invariant strategy combination of the FSG representation. The translation of the frames' language into German results in the attribute Farbe and the label set {grün, braun, ocker}. In contrast to the English frame where all labels consist of the same number of letters, the German label for the green cube is shorter than the others. Therefore, one could argue that Germans should take the green cube. This, however, would violate the language invariance requirement. Thus, language invariance does not account for the inner structure of the individual labels. Of course, one could add an additional attribute that covers the inner structure of the color label, e.g., length_of_color_label. But in this case, the game framed in English would not be isomorphic to the game framed in German.

Suppose now that green is usually regarded as a striking color. Intuitively, the green cube should be chosen by the players. But unless, for example, the property of being striking is explicitly modelled with a separate attribute as striking, the connotation of green as a striking color does not influence the solution.

Reconsider Example 2.2.1 (p. 10). Alternatively, the perception of M_3 by players that realize both color and shape can be described by the frame F with $\mathfrak{A} = \{\text{appearance}\}$, $\Lambda = \{\text{light\&round}, \text{light\&cubic}, \text{dark\&round}\}$, and $\ell_{i\text{appearance}}(s_{i1}) = \text{light\&round}$, $\ell_{i\text{appearance}}(s_{i2}) = \text{light\&cubic}$, $\ell_{i\text{appearance}}(s_{i3}) = \text{dark\&round}$ for both i. In contrast to the findings above (see pp. 43), all strategy-indices are symmetric now. Clearly, it matters how a frame is made up. In both cases, the strategies are distinguished by the labelling but the frame in Subsection 2.5.1 has a richer structure. Hence, it is the structure of the whole labelling rather than the single label that influences the symmetry of strategy-indices.

As the examples above show, it is an extraterrestrial's perspective that is adopted: Properties of strategies that can be recognized directly by terrestrial players and game theorists are considered to be observable by technical devices only and in a very abstract way, similar to frequency plots of the ultrasonic voices of bats. Thus, other things (the payoff structure) being equal, no property and no property label has a meaning or an exploitable structure for these extraterrestrials. In some sense, language invariance steers a course between complete ignorance of the players' description of a game as in traditional game theory on the one hand and full exploitation of this description in the strong sense that strategies with different label vectors cannot be symmetric. This way, language invariance does not simply recognize differences in the strategy descriptions but also accounts for more subtle similarities (see Subsection 2.5.6, pp. 49).

2.5.5 Separating FSGs

In contrast to Bacharach's (1991, 1993) approach, frames may assign the same label to different attributes. This raises the question whether one could do without this feature of FSGs. We are thus led to define separating frames:

Definition 2.5.8 (Separating Frames and FSGs).
Let $(G, F) \in \mathcal{F}$ be the FSG $((I, (S_i)_{i\in I}, (u_i)_{i\in I}), (\mathfrak{A}, \Lambda, (\ell_{ia})_{(i,a)\in I\times\mathfrak{A}}))$. For all $a \in \mathfrak{A}$ define $\Lambda(a)$ as the set of all labels assigned with respect to a,

$$\Lambda(a) := \{\lambda | \exists i \in I, s_i \in S_i : \lambda = \ell_{ia}(s_i)\}. \tag{2.16}$$

Both the frame F and (G, F) are called separating, if $\Lambda(a) \cap \Lambda(a') = \emptyset$ for all $a, a' \in \mathfrak{A}$ with $a \neq a'$.

(a)

s_{1j}	s_{11}	s_{12}
a $\ell_{1a}(s_{1j})$		
front	red	gray
back	red	pink

(b)

s_{1j}	s_{11}	s_{12}
a $\bar{\ell}_{1a}(s_{1j})$		
front	(front, red)	(front, gray)
back	(back, red)	(back, pink)

Fig. 2.16. Transforming a frame into a separating one

Now, the question above can be put more precisely: Is it possible to transform any frame of a strategic game into a separating frame such that as much as possible of the structure of the original frame is kept and that the strategy-index symmetry relation is not affected?

Any frame $F = (\mathfrak{A}, \Lambda, (\ell_{ia})_{(i,a)\in I\times\mathfrak{A}})$ of $G = (I, (S_i)_{i\in I}, (u_i)_{i\in I})$ can be transformed into another frame $F' = (\mathfrak{A}, \Lambda', (\ell'_{ia})_{(i,a)\in I\times\mathfrak{A}})$ of G where $\Lambda' := \mathfrak{A} \times \Lambda$ and $\ell'_{ia}(s_i) := (a, \ell_{ia}(s_i))$ for all $i \in I$, $s_i \in S_i$, and $a \in \mathfrak{A}$. I.e., any label λ of F is split into a collection of labels $((a, \lambda))_{a\in\mathfrak{A}}$ such that any (a, λ) is, if at all, assigned with regard to attribute a only. In a sense, F' keeps as much as possible of the structure of F: for all $i, i' \in I$, $s_i \in S_i$, $s_{i'} \in S_{i'}$, $a \in \mathfrak{A}$, we have $\ell_{ia}(s_i) = \ell_{i'a}(s_{i'})$ if and only if $\ell'_{ia}(s_i) = \ell'_{i'a}(s_{i'})$.

It easy to check that any symmetry $(\mathbf{f}, \mathbf{t}) = ((\pi, (r_i)_{i\in I}), (\mu, \tau))$ of (G, F) induces a symmetry $((\pi, (r_i)_{i\in I}), (\mu, \tau'))$ of (G, F') where $\tau'(a, \lambda) := (\mu(a), \tau(\lambda))$ for all $(a, \lambda) \in \Lambda'$. Thus, symmetric strategy-indices in (G, F) remain symmetric in (G, F'). The converse, however, does not hold; as the following example shows, this transformation might create new symmetry relations between strategy-indices.

Consider a game with one player only who has two strategies between which he is indifferent: (a) to take a book both the front and back of which are red or (b) to take a book the front of which is gray and the back of which is pink. This game can be represented by the FSG (G, F) with $I = \{1\}$, $S_1 = \{s_{11}, s_{12}\}$, $u_1(s_{11}) = u_1(s_{12}) = 1$, $\mathfrak{A} = \{\text{front}, \text{back}\}$, $\Lambda = \{\text{red}, \text{gray}, \text{pink}\}$, and the label functions as in Figure 2.16 (a). Using Corollary 2.5.1, one can show that s_{11} and s_{12} are not symmetric.

Consider now the transformed frame $\bar{F}$ (see Figure 2.16 (b)). Clearly, both strategies are symmetric now. The reason for this is the following: Transforming F into $\bar{F}$ moves some structural features of the frame as a whole into the labels' structure. Since the inner structure of the labels is ignored by isomorphisms and symmetries (see Subsection 2.5.4) new symmetries might arise. The original frame allows the player to say: 'I take the book of which both the front and the back have the same color.' This is impossible with the separating frame $\bar{F}$; now, front and back of the books are 'colored' differently by definition.

2.5.6 Applications

This subsection presents some first applications and implications of the FSG approach: The first subsubsection compares language invariance in FSGs with the principle of insufficient reason employed in other formalizations of static focal points. In the second subsubsection, we demonstrate how the choice of a salient option can be justified within the FSG framework. The third subsubsection explains the subtle salience in Example 2.2.1 (p. 10) within the FSG approach. The use of player labels is demonstrated in the fourth subsubsection. The final subsubsection illustrates that the FSG approach cannot explain focal points in bargaining.

Language Invariance and the Principle of Insufficient Reason. Symmetry invariance in FSGs, roughly speaking, says that strategies that are indistinguishable—considering the players' description—should obtain the same probabilities. Taken for itself, symmetry invariance can be viewed as some version of the principle of insufficient reason. Since language invariance, upon which symmetry invariance rests, is a more general concept, on the other hand, we can think of it as a way of justifying this principle.

Bacharach (1991), Janssen (1997), Bacharach & Bernasconi (1997), and Bacharach & Stahl (2000) apply some or another version of the principle of insufficient reason. By making up the perceived-option sets, in fact, they base their solution concepts on the partitions of the players' strategy-index sets that are induced by single attributes, i.e. partitions the cells of which contain exactly those of a player's strategy-indices that are assigned the same label with regard to the attribute under consideration (see Subsection 2.4.2). Even this can be interpreted as some expression of this principle. As a consequence thereof, these concepts are language invariant.

What all of these approaches have in common is that they do not exploit the whole structure of the labelling: Janssen compresses all of the players' attribute partitions into a single one before he applies a symmetry requirement upon groups of strategies (see (2.13), p. 2.13). While Janssen considers differences in the whole label vectors in doing so, Bacharach and Bacharach & Bernasconi are concerned with differences along single attributes (see (2.12)). For each player, they introduce an option set as follows: Besides randomization, any label that is actually assigned to a strategy-index with regard to some attribute constitutes an option, namely, randomization over all strategy-indices that obtain this label regarding this attribute. A symmetry disqualification requirement then excludes as a solution any option for which there is another option of the same attribute that covers the same number of strategy-indices. Bacharach & Stahl extend this option set by combining labels of different attributes. However, some of the structural properties of the frame get lost. Nevertheless, in some interesting cases (e.g., the FSG in Subsubsection 2.5.6 for $n > 2$), these concepts and language invariance (together with payoff dominance) point to the same (unique) solution.

Rationalizing the Choice of a Salient Option. The perception of salience in matching games M_n can be modelled by a frame F_n straightforwardly. The language part simply consists of $\mathfrak{A}_S = \{\texttt{salience}\}$ and $\Lambda_S = \{\texttt{salient}, \texttt{non-salient}\}$. Without loss of generality, the label functions of both player-indices i can be set to $\ell^S_{i\texttt{salience}}(s_{i1}) = \texttt{salient}$ and $\ell^S_{i\texttt{salience}}(s_{ij}) = \texttt{non-salient}$ for $j \neq 1$.

Obviously, for $n \geq 3$, on the one hand, the strategy-indices s_{ij} for $j \neq 1$ are symmetric to each other. On the other hand, the strategies s_{11} and s_{21} are also symmetric. However, the strategy-indices in each of the afore mentioned groups are not symmetric to the strategy-indices in the other group. This is so because a symmetry of (M_n, F_n) that constitutes the symmetry of s_{11} and s_{12} would have to map the label $\texttt{salient}$ to the label $\texttt{non-salient}$ and vice versa. Since the labels are assigned to a different number of strategy-indices, using Corollary 2.5.1, this is impossible. Thus, in every symmetry invariant strategy-index, all pure strategy-indices except for s_{i1} have to be assigned the same probabilities. The unique payoff dominant symmetry invariant equilibrium then is the choice of the salient option by both players. So far, the FSG framework leads to the same results as Gauthier (1975).

Things turn out to be different for $n = 2$. In this case, both labels are assigned to just two strategy-indices—to one strategy-index of each player. Therefore, both of a player's strategy-indices could be—and actually are—symmetric. Hence, the unique symmetry invariant strategy-index combination prescribes randomization. This departure from Gauthier is fully in line with the criticism raised by Provis (1977), Gilbert (1989), and Miller (1991)—in a sense, the single non-salient option is as salient as the salient one, and as both players taking the non-salient option yields the same payoff as taking the salient one, both options should be treated in the same way.

Since Bacharach (1991) admits non-exhausting families, i.e. families that do not contain concepts for all strategies, he is able to support the choice of the salient option in the two-option case. As mentioned in Subsection 2.5.1, non-exhausting families (attributes) can be formalized by adding and assigning a label $\texttt{no\_label\_assigned}$. However, the resulting FSG would be isomorphic to the FSG (M_2, F_2), and the choice of the salient option could not be supported. In view of the criticism above, this seems to be not too restrictive.

Subtle Salience. Reconsider Example 2.2.1. For $\pi = \text{id}$, the symmetry of the FSG representation shown in Subsection 2.5.3 establishes the symmetry of s_{i2} and s_{i3} in the FSG representation. However, there is no symmetry that maps s_{i2} or s_{i3} to s_{i1}. If such a symmetry existed, one of the labels $\texttt{dark}$ or $\texttt{cubic}$ would have to be mapped to one of the labels $\texttt{light}$ or $\texttt{round}$. Since the first two labels are assigned to less strategy-attribute pairs than the latter, this is in contradiction with Corollary 2.5.1. Moreover, it easy to see that both players' strategies related to the same object are symmetric. Thus, symmetry invariance requires the players to take (a) the same strategy where (b) the light cube and the dark ball get the same probabilities. Choosing the light ball—the salient option—is the unique symmetry invariant equilibrium of this FSG that is not dominated by another symmetry invariant equilibrium.

This example shows that the introduced notion of symmetry invariance does not simply account for differences between the descriptions of the strategies, but is able to detect more subtle similarities within the structure of the labelling. While Janssen has to make some additional assumptions (see Section 2.2), the FSG approach is able to derive the subtle salience from the given description of the game only. In doing so, symmetry invariance together with payoff dominance endogenizes salience to some extent.

Labelling Players. The following example demonstrates how labels that refer to properties of players can be used within the FSG framework. Consider the following game.

Example 2.5.3 (Take Two Black Balls).
Three players—two women and a man—are each given a basket containing one white ball and one black ball. In separate rooms and without the possibility of communication, the players take one ball from their basket and hand it to the game master. If exactly two black balls have been chosen, all players get a prize, otherwise they get nothing.

Both strategies of the three players are distinguished by the payoff structure, but the players themselves are not—the players are symmetric in the sense of Definition 2.3.11. However, the players' problem is to find out two of them to take the black ball. Intuitively, it is clear that both women should choose the black ball. This intuition can be supported within the FSG framework. Figure 2.17 shows a strategic game for this game. Player 1 is row chooser, player 2 column chooser, and player 3 matrix chooser. In each cell, the bottom left number is player 1's payoff, the number in the middle is player 2's payoff, and the top right number is player 3's payoff. A frame for players that realize the sex of players is F such that $\Lambda = \{\mathtt{female}, \mathtt{male}\}$, $\mathfrak{A} = \{\mathtt{sex}\}$ and $\ell_{i\mathtt{sex}}(s_{ij}) = \mathtt{female}$ for $i \in \{1, 2\}$ and $\ell_{3\mathtt{sex}}(s_{3j}) = \mathtt{male}$.

According to Corollary 2.5.1 the labels cannot be symmetric. Therefore, the male player is not symmetric to any of the female players, but the female players are symmetric to each other. Symmetry invariance implies that both women take the 'same' strategy. Payoff dominance then selects the intuitive solution—both women take the black ball, and the man takes the white ball.

Language Invariance and Bargaining. The FSG approach is primarily intended to break symmetries. Therefore, as argued in Subsection 1.1.2, focal points in bargaining games cannot be supported within this framework.

Reconsider the bargaining game from Example 1.1.1. The arbitrator's suggestion of the splitting $(3, 2)$ can be represented by a framed version of the original game (see Figure 1.4) as follows: The attribute set $\mathfrak{A}$ contains the single attribute $\mathtt{suggested}$, the label set Λ comprises the labels $\mathtt{yes}$ and $\mathtt{no}$, and the label functions are set as in Figure 2.18, i.e., the strategy-indices

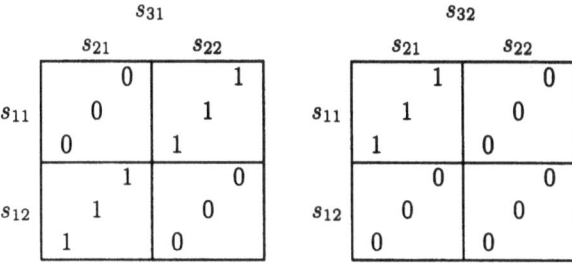

Fig. 2.17. Take Two Black Balls

a	s_{ij} $\ell_{1a}(s_{ij})$	s_{11}	s_{12}	s_{13}	s_{14}	s_{21}	s_{22}	s_{23}	s_{24}
suggested		no	no	yes	no	no	yes	no	no

Fig. 2.18. Labelling of a simple bargaining game according to an arbitrator's suggestion

corresponding to the arbitrator's suggested division get the label **yes** and all other strategy-indices get the label **no**.

In the underlying strategic game, the strategies s_{1k} and s_{2k} can be symmetric only. However, in the framed version, even this is impossible— Corollary 2.5.1 prevents the symmetry of the strategy-indices s_{13} and s_{23}, hence the symmetry of the players. This way, the FSG representation of the arbitrator's suggestion supports our intuition that a focal division might break the symmetry of the players, but again, arguing along the lines of Subsection 1.1.2, this does not support the suggested division as the focal point to be chosen.

2.6 Focal Points in FSG Representations of VUMGs

Focal points in VUMGs as considered in Subsection 2.4.2 can be formalized within the FSG approach. In order to do this, FSG representations of VUMGs as well as focal variable universe equilibria of these games are introduced in the first subsection. The second subsection reconsiders Example 2.2.2 within this framework. Introducing a reduction algorithm for frames, the final subsection shows that Janssen's (1997) results (see Theorem 2.4.1, p. 32) also hold for the FSG representations of VUMGs.

2.6.1 FSG Representations of VUMGs

Similar to the strategic game representation (see pp. 29), a VUMG $(M_n, \mathfrak{A}, (\ell_{ia}), v)$ can be represented by a set of FSGs $(G_r, F_r)_{r \in \mathbf{R}}$. For all repertoires r, G_r is the same as M_n except for the payoff functions. Let $\theta \in \Theta := \bigtimes_{r \in \mathbf{R}} \Sigma$ be a vector containing mixed-strategy combinations for each G_r. Given $\theta \in \Theta$, the payoff function in G_r is defined by

$$u_i^r(s) := \sum_{r' \subset r} v(r'|r) u_i(s_i, \theta_{r'-i}) + v(r|r) u_i(s) \tag{2.17}$$

for all $s \in S$ and $i \in I$ where $-i$ denotes i's opponent and $\theta_{r'-i}$ denotes $-i$'s strategy in θ for $G_{r'}$. I.e., u_i^r is the expected payoff of player i under the assumption that the subrepertoire players follow θ. The frame $F_r := (r, \bigcup_{a \in r} a, (\ell_{ia})_{(i,a) \in I \times r})$ contains the repertoire r as attribute set, the union of r's attributes as label set, and the appropriate label functions ℓ_{ia} the range of which is extended to $\bigcup_{a \in r} a$.

Definition 2.6.1 (Focal variable universe equilibria of VUMGs).
A vector $\theta \in \Theta$ is called a focal variable universe equilibrium (FVUE) of the VUMG $(M_n, \mathfrak{A}, (\ell_{ia}), v)$ if, for all r, $\theta_r \in \Sigma$ is the unique symmetry invariant equilibrium of (G_r, F_r) that is not dominated by another symmetry invariant equilibrium.

The difference between strategic game representations and FSG representations of VUMGs lies in the way they model and exploit the players' labelling of the game. The strategic game representations model the labelling and restrict the repertoire agents' strategy sets by forming perceived-option sets simultaneously. In contrast, modelling and restricting are separate things in the FSG representation—frames model the labelling and the symmetry invariance requirement actually restricts the players' strategic possibilities.

2.6.2 An Example

Reconsider the game from Example 2.2.2 (p. 12). This game can be modelled as a FSG representation of a VUMG. Assuming $v(\{\texttt{color}\}) = 1 - \bar{v}$ and $v(\{\texttt{color}, \texttt{grain}\}) = \bar{v}$ for some $\bar{v} \in [0,1]$, two repertoires have to be considered only—$\{\texttt{color}\}$ and $\{\texttt{color}, \texttt{grain}\}$. In $G_{\{\texttt{color}\}}$, the labelling is such that $\ell_{i\texttt{color}}(s_{i1}) = \ell_{i\texttt{color}}(s_{i2}) = \texttt{red}$ and $\ell_{i\texttt{color}}(s_{ik}) = \texttt{yellow}$ for $k \in \{3, \ldots, 20\}$ and $i \in I$. Additionally, we have $\ell_{i\texttt{grain}}(s_{i3}) = \texttt{wavy}$ and $\ell_{i\texttt{grain}}(s_{ik}) = \texttt{straight}$ for $i \in I$ in $G_{\{\texttt{color}, \texttt{grain}\}}$. Obviously, randomizing over the red blocks is the $G_{\{\texttt{color}\}}$-part in a FVUE. Thus, we have the payoff matrix for player 1 in $G_{\{\texttt{color}, \texttt{grain}\}}$ as in Figure 2.19.

By Corollary 2.5.1 the labels are not symmetric. Thus, the strategies within the groups $\{s_{i1}, s_{i2}\}$, $\{s_{13}, s_{23}\}$, and $\{s_{i4}, \ldots, s_{i20}\}$ only are symmetric. As in the VUMG, for $\bar{v} > \frac{1}{2}$, taking the wavy yellow block, and for $\bar{v} < \frac{1}{2}$, picking a red block is the $G_{\{\texttt{color}, \texttt{grain}\}}$-part of the FVUE.

	s_{21}	s_{22}	s_{23}	s_{24}	$\cdots$	s_{220}
s_{11}	$\frac{1}{2}(1-\bar{v})+\bar{v}$	$\frac{1}{2}(1-\bar{v})$	$\frac{1}{2}(1-\bar{v})$	$\frac{1}{2}(1-\bar{v})$	$\cdots$	$\frac{1}{2}(1-\bar{v})$
s_{12}	$\frac{1}{2}(1-\bar{v})$	$\frac{1}{2}(1-\bar{v})+\bar{v}$	$\frac{1}{2}(1-\bar{v})$	$\frac{1}{2}(1-\bar{v})$	$\cdots$	$\frac{1}{2}(1-\bar{v})$
s_{13}	0	0	$\bar{v}$	0	$\cdots$	0
s_{14}	0	0	0	$\bar{v}$	$\cdots$	0
$\vdots$	$\vdots$	$\vdots$	$\vdots$	$\vdots$	$\ddots$	0
s_{120}	0	0	0	0	0	$\bar{v}$

Fig. 2.19. Payoff matrix for $G_{\{\text{color,grain}\}}$

2.6.3 Reduction of FSGs and Focal Points in VUMGs

Many of the FSG representations of VUMGs lack FVUE because of the uniqueness requirement in Definition 2.6.1. In order to remedy a similar shortcoming, Janssen (1997) suggests a reduction algorithm for the strategy sets considered by the players depending on the set of attributes they realize (see Definition 2.4.4, p. 32). Inspired by this mechanism, we introduce a reduction algorithm for FSGs.

Definition 2.6.2 (Reduced forms of FSGs).
The reduced form $\mathrm{red}(G,F)$ *of* $(G,F) \in \mathcal{F}$ *is the FSG* $(G,F') \in \mathcal{F}$ *with* $F' = (\mathfrak{A}', \Lambda', (\ell'_{ia}))$ *such that*

1. $\mathfrak{A}' = \{\mathrm{sym}\}$,
2. Λ' *is the set of equivalence classes of* $\bigcup_{i \in I} S_i$ *with respect to the strategy-symmetry relation of Definition 2.5.7, and*
3. $s_i \in \ell_{i\,\mathrm{sym}}(s_i)$ *for all* $i \in I$ *and* $s_i \in S_i$.

After reduction, the frame of a FSG reflects the symmetry relation between the strategy-indices only. Obviously, the reduction cannot destroy the symmetry of strategy-indices—strategy-indices that have been symmetric before a reduction remain symmetric after the reduction of the frame. Thus, we have:

Theorem 2.6.1.
If $\sigma \in \Sigma$ *is symmetry invariant in* $\mathrm{red}(G,F)$ *then it is also symmetry invariant in* (G,F).

This reduction can be applied repeatedly to the resulting FSGs. Except for the initial reduction, each following reduction step does not increase the number of labels of the frame. As the number of labels is finite, the reduction process eventually becomes stationary after a finite number of steps.

Definition 2.6.3 (Fully reduced forms of FSGs).
The fully reduced form of $(G,F) \in \mathcal{F}$ *is the FSG* $\mathrm{red}^*(G,F) \in \mathcal{F}$ *such that*

1. $\mathrm{red}^*(G, F) = \mathrm{red}^n(G, F)$ *for some* $n \in \mathbb{N}$ *and*
2. $\mathrm{red}^n(G, F) = \mathrm{red}^{n+1}(G, F)$.

Note that $\mathrm{red}^*(G, F)$ does not depend on the choice of n and therefore is always well-defined.

Definition 2.6.4 (Focal points of VUMGs).
A vector $\theta \in \Theta$ is called a focal point of the VUMG $(M_n, \mathfrak{A}, (\ell_{ia}), v)$ if for all r, $\theta_r \in \Sigma$ is the unique symmetry invariant strategy-index combination of $\mathrm{red}^*(G_r, F_r)$ that is not dominated by another symmetry invariant strategy-index combination.

The following theorem shows that similar results as those of Theorem 2.4.1 (p. 32) hold for FSG representations of VUMGs. The proof can easily be adapted from Casajus (1997). Again, this theorem explains the phenomenon that players coordinate much more frequently in matching games than by randomizing.

Theorem 2.6.2.
Let be $(M_n, \mathfrak{A}, (\ell_{ia}), v) \in G(n, \mathfrak{A}, (\ell_{ia}))$ a VUMG and $(G_r, F_r)_{r \in \mathbf{R}}$ its FSG representation according to Subsection 2.6.1.

1. *In generic cases, $(M_n, \mathfrak{A}, (\ell_{ia}), v)$ has a focal point $\theta^* \in \Theta$.*
2. *If θ^* is a focal point of $(M_n, \mathfrak{A}, (\ell_{ia}), v)$ then θ_r^* is an equilibrium of (G_r, F_r) for all $r \in \mathbf{R}$.*
3. *In generic cases, a focal point θ^* of $(M_n, \mathfrak{A}, (\ell_{ia}), v)$ has the following property: If $a \in r$ and $u_i^{\{a\}}(\theta_{\{a\}}^*) > \frac{1}{n}$ then $u_i^r(\theta_r^*) > \frac{1}{n}$ for all $r \in \mathbf{R}$ and $a \in \mathfrak{A}$.*

3 Weak Isomorphisms of Extensive Games

While there is an established notion of isomorphisms of strategic games (see Subsection 2.3.4), there is no such notion for extensive games. Mainly, the reason for this gap seems to be the more complex nature of extensive games in connection with the resulting ambiguities in interpreting them. In this chapter, weak isomorphisms of extensive games are introduced and advocated.

3.1 Introduction

As Harsanyi & Selten's (1988) isomorphisms of strategic games, isomorphisms of extensive games can be viewed as a means to identify *structurally* similar extensive games and to identify corresponding structural elements of these games—players, information sets, actions, and nodes. And it is this emphasis of structural features that distinguishes isomorphisms from considerations of strategic equivalence as Kohlberg & Mertens' (1986) invariance requirement or Thompson's (1952) and Elmes & Reny's (1994) transformations. Concerning strategic games, isomorphisms are bijective mappings between the players' pure-strategy sets that preserve the player structure and the payoff structure. In extensive games, basically, these mappings can be based either on the action partitions or on the node sets. Actually, both approaches have been adopted in the literature. In a sense, the traditional tree representation of extensive games (see e.g. Selten 1975) corresponds to isomorphisms that are based on bijections of the node sets, whereas the sequence representation (see e.g. Osborne & Rubinstein 1994) corresponds to isomorphisms based on bijections of the action partitions.

Since strategic games have a simple structure, there seems to be much agreement that Harsanyi & Selten's (1988) concept of isomorphisms captures their structural equivalence. Owing to the sequential nature of moves, in contrast, extensive games have a much richer structure. Therefore, there are (more) ambiguities in interpreting them, a fact which also has impact on the notion of isomorphisms of extensive games. One ambiguity concerns the sequence of moves: As a technical feature of extensive games, the players have to move in some order. Thus, simultaneous moves as in the traditional extensive representation of strategic games (see Subsection 3.5.1), are modelled as sequential ones, but such that the players have no information about the moves of the preceding players. I.e., the sequence of moves in these games should not be interpreted as a real one. Consequently, one could argue that games which differ in that inessential kind of the order of moves should be isomorphic. Following this argument, the crucial question is: To what extent is the sequence of moves inessential? Or, to put it differently: When does an isomorphism respect the order of moves to the essential extent?

The following criterion is suggested: A concept of isomorphisms respects the order of moves to the essential extent if its isomorphisms always carry equilibria of the kind under consideration into equilibria of the same kind, i.e., if the equilibrium concept is invariant with respect to these isomorphisms (see Subsection 3.3.1). Of course, this property is especially important with respect to those equilibrium concepts that explicitly refer to the sequential nature of moves as the concepts of subgame perfect equilibria (Selten 1975) or sequential equilibria (Kreps & Wilson 1982). Beyond that, the order of moves seems to be rather inessential, i.e., it constitutes a pure labelling of players only and therefore should not affect isomorphisms. Note that this criterion reverses the relation between equilibrium/solution concepts and invariance with respect to isomorphisms: Usually, invariance with respect to isomorphisms serves as a measure to assess solution concepts or as a requirement on solution concepts. Here, in contrast, the solution concepts are exploited to assess some concept of isomorphisms. Of course, this reversal is based on the assumption that the equilibrium concepts used for this kind of assessment are sound in the sense that they do not contradict our intuitions.

Like Elmes & Reny (1994), Peleg, Rosenmüller & Sudhölter (1999) introduce (strong) isomorphisms of extensive games that rest upon bijections of the node sets which respect the order of moves in a very strong sense. As it turns out, these strong isomorphisms are incompatible with the traditional extensive representations of strategic games (see Subsection 3.5.1)—symmetric strategies in a strategic game may not be symmetric in the extensive representation. Thus, one could argue that Peleg et al.'s (1999) strong isomorphisms are *too* strong. In order to remedy this incompatibility, Peleg, Rosenmüller & Sudhölter (2000) introduce a new kind of extensive representation—*canonical extensive forms of game forms*—for which both notions of symmetry coincide. Besides their axiomatic foundation, the idea is quite simple: (a) An initial chance move selects any traditional representation—in which the players move in some order—with the same probability, and (b) since the players are assumed to be not informed about this chance move, the player (information, action) partitions of all traditional representations are to be connected in the straightforward way. Obviously, these representations are considerably more complex and more difficult to deal with than the traditional ones. So a notion of isomorphisms of extensive games that fits the traditional representation of strategic games seems to be desirable.

In contrast, Selten (1983), Oh (1995), and Casajus (1998) base their (weak) symmetries of extensive games on bijective mappings of the action partitions. In addition, these (weak) symmetries do not respect the order of moves in the very strict sense of Peleg et al. (1999). This way, their symmetries are fully compatible with the traditional extensive representations of strategic games (Casajus 1998, Theorem 4.8). Additionally, in non-pathological cases, weak symmetries are subgame preserving (Selten 1983, Theorem 1). Also, (weakly) symmetry invariant equilibria do always exist (Casajus 1998, Theorem 4.9). Beyond these appealing properties, however, no further justification for weak symmetries is given. So it is not clear, whether weak symmetries respect the order of moves in the essential extent as suggested above. Of course, subgame preservation goes some way to answer this question in the affirmative.

Besides the complete games, Peleg et al. (1999) consider what they call preforms and game forms. While the latter, in principle, are games without payoff functions but introduce some kind of pure labelling of the results of the players' actions. 'Pure', since the outcomes that are assigned to the strategy combinations/terminal nodes are not related to a preference order. Note that this kind of labelling may restrict the possible payoff assignments to terminal nodes. In contrast, we consider isomorphisms of complete games only. The respective notions for preforms can easily be obtained by dropping the payoff functions and the payoff-related transformation/condition in the definition of weak isomorphisms. Concerning preforms, we feel that the pure labelling—of outcomes as well as of players and strategies/actions—should be separated from the strategic description of a game, i.e., it should be put into a separate structure as the frames introduced in the Sections 2.5 and 4.4.

This chapter extends Casajus' (1998) symmetries into weak isomorphisms of extensive games in a straightforward way and explores some of their properties. It is shown that weak isomorphisms are just "weak" enough to disregard the irrelevant order of moves on the one hand, but without ignoring the strategically relevant order of moves and without violating essential sequential rationality considerations on the other hand. So weak isomorphisms are an adequate means to describe structural similarities of extensive games *if* one views the sequential nature of moves as a technical peculiarity of extensive games rather than a representation of real world facts. Of course, this does not mean that sequential moves in a game never correspond to sequential actions in real life.

The outline of the chapter is as follows: The next section gives basic definitions and notation concerning finite extensive games with perfect recall. In the third section, weak isomorphisms of extensive games and the requirement of invariance with respect to isomorphisms upon their solutions are introduced and discussed. Subsections explore some of their properties: Weak symmetries, weakly symmetric actions, and weak symmetry invariance are introduced, and some results concerning the existence of weakly symmetry invariant equilibria are established. Also, results related to the preservation of

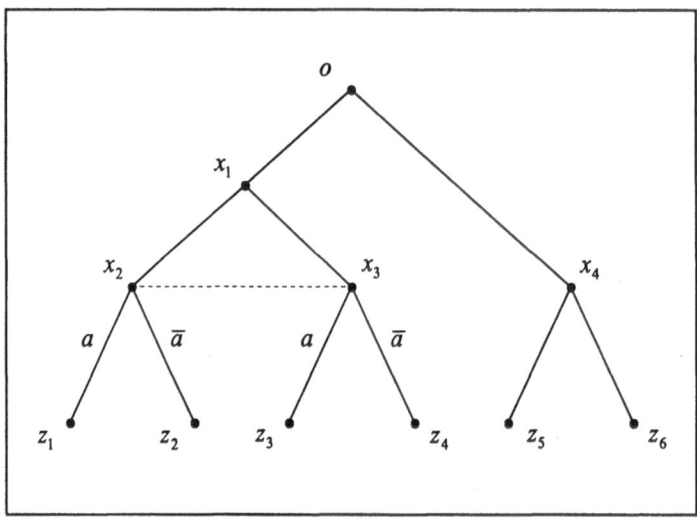

Fig. 3.1. An extensive game

the relevant aspects of the order of moves are presented. In the fourth section, the relation between weak isomorphisms and Peleg et al.'s (1999) strong isomorphisms is explored. The compatibility of strategy symmetry in strategic games and weak strategy symmetry in their traditional extensive representations is established in the fifth section. The final section summarizes the main results of this chapter.

3.2 Finite Extensive Games with Perfect Recall

In this chapter, we only consider finite extensive games with perfect recall; the class of these games is denoted by $\mathcal{G}^e$. Throughout, we refer to these games as extensive games. The definitions given below are very similar to those given by Selten (1975).

3.2.1 Definition

A finite extensive game Γ is an octuplet $(T, \lhd, I, P, H, A, p, u)$ where its constituents are defined below. These definitions are illustrated by help of the game $\bar{\Gamma}$ in Figure 3.1.

The non-empty and finite set of nodes T and the binary relation $\lhd$ on T—'precedes' or 'is succeeded by'—constitute a tree. I.e., $(T, \lhd)$ has the following properties: The node set T contains at least two nodes. The relation $\lhd$ is (a) transitive and (b) asymmetric. (c) There is a node o—the root—such that $o \lhd x$ for all nodes $x \neq o$. (d) For arbitrary $x, x', x'' \in T$, $x' \lhd x$, $x'' \lhd x$ and $x' \neq x''$ imply $x' \lhd x''$ or $x'' \lhd x'$.

Nodes that are not succeeded by another node are called terminal nodes. The set of terminal nodes is denoted by Z; the set of non-terminal nodes—decision nodes—is denoted by X. The setting above secures that each node x except for the root has exactly one immediate predecessor $V(x)$; $N(x)$ denotes the set of all of x's immediate successors. The path $\psi(x)$ of a node x is a set containing x itself and all preceding nodes, except for the root—$\psi(o) = \emptyset$ and $\psi(x) = \psi(V(x)) \cup \{x\}$. The ordered path $\psi(x)$ of a node x is the vector $(\psi_k(x))_{k \in \{1,\ldots,|\psi(x)|\}}$ such that $\psi_k(x) \in T$, $\psi_{|\psi(x)|}(x) = x$, and $\psi_{k-1}(x) = V(\psi_k(x))$ for all $1 < k \le |\psi(x)|$. Hence, $V(\psi_1(x)) = o$ for all $x \in T \backslash \{o\}$. An (ordered) play is an (ordered) path of a terminal node.

Example 3.2.1.
In $\bar{\Gamma}$, we have $T = \{o, x_1, x_2, x_3, x_4, z_1, z_2, z_3, z_4, z_5, z_6\}$ and, for example, $o \lhd x_2$, $x_1 \lhd x_2$, $x_3 \lhd z_4$, but $x_1 \ntriangleleft z_6$; $Z = \{z_1, z_2, z_3, z_4, z_5, z_6\}$ and $X = \{o, x_1, x_2, x_3, x_4\}$. For example, we also have $N(o) = \{x_1, x_4\}$, $N(x_3) = \{z_3, z_4\}$, $\psi(z_5) = \{x_4, z_5\}$, and $\psi(x_3) = (\psi_1(x_3), \psi_2(x_3)) = (x_1, x_3)$.

The *player set* I contains the players $i, i', \ldots, i^{'\ldots'}$ and one distinguished player i_0 who represents the chance mechanism. The players $I \backslash i_0$ are called personal players.

The *player partition* P partitions X into player cells P_i—sets of decision nodes that are controlled by the same decision maker. While the player cells P_i of the personal players $i \in I \backslash i_0$ are always non-empty, i_0's cell may be empty. Each decision node x is assigned the player $i(x)$ who controls it by the player function $i : X \to I$, where $i(x) := i'$ if $x \in P_{i'}$ for all $i' \in I$ and $x \in X$.

Example 3.2.2.
In $\bar{\Gamma}$, let I be $\{1, 2\}$ and $P_1 = \{o, x_2, x_3, x_4\}$ and $P_2 = \{x_1\}$. I.e., there are two players—1 and 2, where player 1 moves at the decision nodes o, x_2, x_3, and x_4; and player 2 moves at the decision node x_2. Note that there is no chance player in $\bar{\Gamma}$; an example for a game with a chance mechanism is given below.

The *information partition* H partitions the set of decision nodes into information sets h; $h(x)$ denotes the information set containing the decision node x. The interpretation of $h(x)$ is that a player who is choosing at x is uncertain whether he is at x or at some other $x' \in h(x)$. The information partition has the following properties: (a) H is a subpartition of P. (b) For all $h \in H$ and $x, x' \in h$, we have $|N(x)| = |N(x')|$. (c) For all $x \in P_{i_0}$, we have $h(x) = \{x\}$. $H_i := h(P_i)$ denotes the set of player i's information sets.

Example 3.2.3.
In $\bar{\Gamma}$, we have $H = \{\{o\}, \{x_1\}, \{x_2, x_3\}, \{x_4\}\}$, $H_1 = \{\{o\}, \{x_2, x_3\}, \{x_4\}\}$, and $H_2 = \{\{x_1\}\}$. I.e., player 2 who has to move at x_1 knows that he moves at the node x_1—$h(x_1) = \{x_1\}$. In contrast, when player 1 has to move at

node x_3, he does not know whether he is at node x_2 or at node x_3—$h(x_3) = \{x_2, x_3\}$, i.e., player 1 does not know how player 2 acted at x_1. In Figure 3.1, the information set $\{x_2, x_3\}$ is indicated by a dashed line.

The *action partition* A partitions the set of non-root nodes into actions a; $a(x)$ denotes the action containing the non-root node x. The interpretation of $a(x)$ is that the nodes in $a(x)$ are reached by the same act. The action partition has the following property: For all $a \in A$, there is a $h \in H$ such that $V(a) = h$ and $|a| = |h|$. $A_h := a(N(h))$ denotes the actions at information set h, and $A_i := a(N(P_i))$ denotes the set of actions of player i.

Example 3.2.4.
In $\bar{\Gamma}$, we have $A_1 = \{\{x_1\}, \{x_4\}, \{z_1, z_3\}, \{z_2, z_4\}\{z_5\}, \{z_6\}\}$, $A_2 = \{\{x_2\}, \{x_3\}\}$. Note that this partition is necessary because there may be information sets containing more than one node. Consider the information set $\{x_2, x_3\}$. Since player 1 does not know whether he is at x_2 or at x_3, the action partition specifies a node to be reached from x_2 as well as from x_3 with regard to any course of action. For example, the action $\{z_1, z_3\}$ can be interpreted as follows: By the same choice of player 1, node z_1 is reached if player 1 is at x_2, and node z_3 is reached if player 1 is at x_3. This is indicated by letters at the edges leading to the respective nodes—a refers to action $\{z_1, z_3\}$ and $\bar{a}$ refers to $\{z_2, z_4\}$. For one-node information sets, any successor of the single node constitutes a separate action. In $\bar{\Gamma}$, the action set of player 1 at his information set $\{o\}$ is $A_{\{o\}} = \{\{x_1\}, \{x_4\}\}$, i.e., moving at $\{o\}$, either the node x_1 or the node x_4 is reached.

Alternatively, a play $\psi(z)$ can be characterized by the set of actions that produces it—$a(\psi(z))$, i.e., the path of the terminal node z expressed by actions. Similarly, ordered plays $\psi(z)$ can be characterized by ordered paths of $z \in Z$ expressed by actions—$a(\psi(z)) = (a(\psi_k(z)))_{k \in \{1, \dots, |\psi(z)|\}}$.

Example 3.2.5.
In $\bar{\Gamma}$, for example, we have $a(\psi(z_1)) = \{\{x_1\}, \{x_2\}, \{z_1, z_3\}\}$ and $a(\psi(z_6)) = (\{x_4\}, \{z_6\})$.

The *chance mechanism* p is a set $\{p_h | h \in H_{i_0}\}$ of probability distributions where p_h is a probability distribution over A_h; $p_h(a)$ denotes the probability that $a \in A_h$ is taken. For all $h \in H_{i_0}$ and $a \in A_h$, we have $p_h(a) > 0$.

Example 3.2.6.
Consider the game in Figure 3.4 (p. 69). We have $P_{i_0} = \{x_1, x_2\}$ and $p = \{p_{\{x_1\}}, p_{\{x_2\}}\}$. For example, $p_{\{x_1\}}(\{z_1\}) = \frac{3}{4}$, i.e., the chance mechanism chooses the action $\{z_1\}$ with probability $\frac{3}{4}$ at its information set $\{x_1\}$.

The *payoff function* $u : Z \to \mathbb{R}^{|I \setminus i_0|}$ assigns a payoff vector $(u_i(z))_{i \in I \setminus i_0}$ to each terminal node z containing the payoff $u_i(z)$ for each personal player i.

Some of the results below hold for a restricted set of games only.

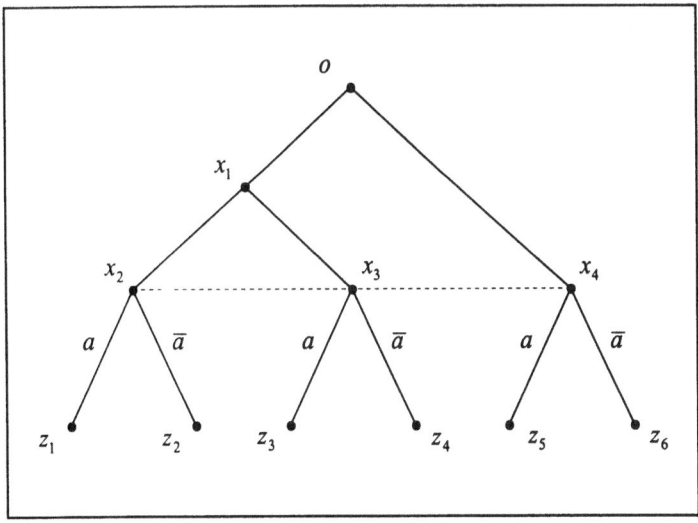

Fig. 3.2. An extensive game with imperfect recall

Definition 3.2.1 (Non-pathological extensive games).

An extensive game is called non-pathological if every decision node is followed by at least two nodes; otherwise it is called pathological.

This implies that the action sets of all information sets contain at least two actions. Since information sets with a single action are of little interest, the focus on non-pathological games does not seem to be too restrictive.

3.2.2 Perfect Recall

Game theorists usually assume that the structure of the game is common knowledge to the players. I.e., all players do not only know the whole structure of the game, but they know that all players know this structure, and they know that all players know that all players know ... and so on ad infinitum (see Lewis 1969, Aumann 1976). In particular, at any *node*, a player knows which actions he has taken before. Virtually always, a stronger assumption is made—*perfect recall.*

Definition 3.2.2 (Perfect recall).

An extensive game exhibits perfect recall, if for all $i \in I \backslash i_0$, $a \in A_i$ and $x \in P_i$, $\psi(x) \cap a \neq \emptyset$ implies $\psi(x') \cap a \neq \emptyset$ for all $x' \in h(x)$.

This can be interpreted as follows: Suppose that a is an action of the personal player $i(x)$ which is on the path of some decision node x $(\psi(x) \cap a \neq \emptyset)$, but which is not on the path of another decision node x' of $i(x)$ $(\psi(x') \cap a = \emptyset)$. If player $i(x)$ remembers his earlier moves, then he should be able to distinguish x and x', i.e., $x' \notin h(x)$. Players never forget their

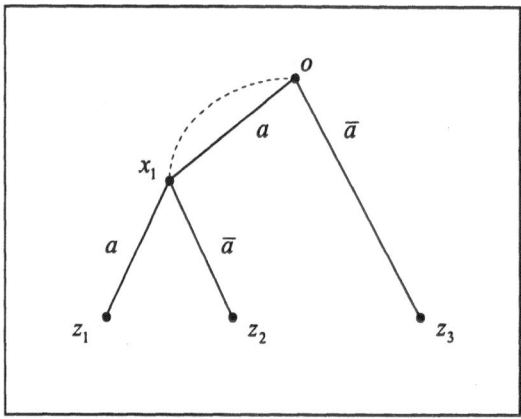

Fig. 3.3. Imperfect recall: absentmindedness

previous moves—at any *information set*, the respective player knows which actions he has taken before.

Example 3.2.7.
Consider the game $\tilde{\Gamma}$ in Figure 3.2 which is quite the same as the game $\bar{\Gamma}$ in Figure 3.1, except for that the information sets $\{x_2, x_3\}$ and $\{x_4\}$ of player 1, and the actions $\{z_1, z_3\}$ and $\{z_5\}$, and $\{z_2, z_4\}$ and $\{z_6\}$ have been merged, respectively. This is indicated by the extended dashed line and the additional letters. Since imperfect recall occurs at information sets with more than one node only, we focus on player 1's information sets $\{x_2, x_3\}$ in $\bar{\Gamma}$ and $\{x_2, x_3, x_4\}$ in $\tilde{\Gamma}$. In $\bar{\Gamma}$, player 1's action $\{x_1\}$ is on the path $\psi(x_2)$ as well as on the path $\psi(x_3)$. Hence, $\bar{\Gamma}$ exhibits perfect recall. Things are different in $\tilde{\Gamma}$. While player 1's action $\{x_4\}$ comes before node x_4, $\{x_4\}$ does not precede x_2. This means that player 1 cannot remember at information set $\{x_2, x_3, x_4\}$ whether he has taken action $\{x_4\}$ or not—$\tilde{\Gamma}$ exhibits imperfect recall.

Piccione & Rubinstein (1997) deal with another kind of imperfect recall—absentmindedness. Consider the game in Figure 3.3. There is one player who moves once at the information set $\{o, x_1\}$ (indicated by the dashed line) and who can either take action $a = \{x_1, z_1\}$ or $\bar{a} = \{z_2, z_3\}$. Interestingly, he cannot remember whether he has already taken a or not. Clearly, this contradicts perfect recall.

Remark 3.2.1.
Perfect recall implies that (a) no action appears more than once within an (ordered) path, i.e., $|a \cap \psi(x)| \leq 1$ for all $x \in T$ and $a \in A$, and that (b) no information set is intersected more than once by the successors of a path, i.e., $|V(\psi(x)) \cap h| \leq 1$ for all $x \in T$ and $h \in H$. Note that the latter is part of Selten's (1975) definition of information sets.

Also, perfect recall implies a one-to-one relationship between plays expressed by actions and terminal nodes: We have to show that different terminal nodes induce different plays expressed by actions. Suppose there were two terminal nodes z and z' with the same path expressed by actions—$a(\psi(z)) = a(\psi(z'))$. Since z and z' are different, there is some node x' where the paths $\psi(z)$ and $\psi(z')$ are branching. Clearly, the successor of x' in $\psi(z)$, x'', belongs to a different action than its successor in $\psi(z')$. Since z and z' have the same path expressed by actions, $a(x'')$ is also in the path of z'. Hence, $h(x)$ is intersected by $V(\psi(z'))$ more than once—at x and at some other $x''' \in V(a(x'')) \cap V(\psi(z'))$. A contradiction.

3.2.3 Strategies

A pure strategy of player $i \in I$ is a vector $\mathbf{a}_i \in \mathbf{A}_i := \bigtimes_{h \in H_i} A_h$; a pure-strategy combination is a vector $\mathbf{a} \in \mathbf{A} := \bigtimes_{h \in H \setminus H_{i_0}} A_h = \bigtimes_{i \in I \setminus i_0} \mathbf{A}_i$ where $\mathbf{a}_i$ also denotes player i's pure strategy in $\mathbf{a}$. An extended pure-strategy combination is a vector $\mathbf{a}^+ \in \mathbf{A}^+ := \mathbf{A} \times \mathbf{A}_{i_0}$.

Each $\mathbf{a}^+ \in \mathbf{A}^+$ is assigned exactly one terminal node $z(\mathbf{a}^+)$ satisfying

$$a(\psi(z(\mathbf{a}^+))) \subseteq \{\mathbf{a}_h^+, h \in H\}, \tag{3.1}$$

where $\mathbf{a}_h^+$ is the action at h in $\mathbf{a}^+$. In view of the definition of an extensive game, it should be clear that such an terminal node exists. By perfect recall and Remark 3.2.1, it is unique. In abuse of notation, let z denote the function $z : \mathbf{A}^+ \to Z$ that assigns this terminal node to an extended action combination. Thereby, the payoff functions can be extended to $\mathbf{A}^+$—$u_i(\mathbf{a}^+) := u_i(z(\mathbf{a}^+))$ for all $i \in I \setminus i_0$ and $\mathbf{a}^+ \in \mathbf{A}^+$.

A mixed strategy of player $i \in I \setminus i_0$ is a vector $\sigma_i \in \Sigma_i$ where Σ_i is the set of probability distributions over $\mathbf{A}_i$, and $\sigma_i(\mathbf{a}_i)$ denotes the probability of the pure strategy $\mathbf{a}_i$. A mixed-strategy combination is a vector $\sigma \in \Sigma := \bigtimes_{i \in I \setminus i_0} \Sigma_i$ where σ_i denotes the mixed strategy of player i in σ. Each mixed strategy combination σ constitutes probability distributions over $\mathbf{A}^+$ and Z with

$$\mathrm{prob}(\mathbf{a}^+|\sigma) = \prod_{h \in H_{i_0}} p_h(\mathbf{a}_h^+) \cdot \prod_{i \in I \setminus i_0} \sigma_i(\mathbf{a}_i) \tag{3.2}$$

$$\mathrm{prob}(z'|\sigma) = \sum_{\mathbf{a}^+ \in \mathbf{A}^+ : z(\mathbf{a}^+) = z'} \mathrm{prob}(\mathbf{a}^+|\sigma) \tag{3.3}$$

for all $\mathbf{a}^+ \in \mathbf{A}^+$ and $z' \in Z$, respectively. By this, the payoff functions can be extended to the domain Σ by $u_i(\sigma) := \sum_{\mathbf{a}^+ \in \mathbf{A}^+} \mathrm{prob}(\mathbf{a}^+|\sigma) u_i(\mathbf{a}^+)$ for all $\sigma \in \Sigma$ and $i \in I \setminus i_0$.

A local strategy of player $i \in I \setminus i_0$ at $h \in H_i$ is a probability distribution b_h over A_h where $b_h(a)$ denotes the probability of $a \in A_h$ in b_h; the set of player i's local strategies at h is denoted by B_h. A behavior strategy of player $i \in I \backslash i_0$ is a vector $b_i = (b_h)_{h \in H_i} \in B_i := \times_{h \in H_i} B_h$ of local strategies for all of his information sets. A behavior-strategy combination is a vector $b \in B := \times_{i \in I \backslash i_0} B_i = \times_{h \in H \backslash H_{i_0}} B_h$; $B^0 \subseteq B$ denotes the set of all completely mixed behavior strategy combinations. Each behavior-strategy combination b constitutes probability distributions over $\mathbf{A}^+$ and Z with

$$\text{prob}(\mathbf{a}^+|b) = \prod_{h \in H_{i_0}} p_h(\mathbf{a}_h^+) \cdot \prod_{h \in H \backslash H_{i_0}} b_h(\mathbf{a}_h^+) \qquad (3.4)$$

$$\text{prob}(z'|b) = \sum_{\mathbf{a}^+ \in \mathbf{A}^+ : z(\mathbf{a}^+) = z'} \text{prob}(\mathbf{a}^+|b) \qquad (3.5)$$

for all $\mathbf{a}^+ \in \mathbf{A}^+$ and $z' \in Z$, respectively. Thereby, the payoff functions can be extended to the domain B and $\mathbf{A}$ by

$$u_i(b) := \sum_{\mathbf{a}^+ \in \mathbf{A}^+} \text{prob}(\mathbf{a}^+|b) u_i(\mathbf{a}^+) \qquad (3.6)$$

for all $b \in B$ and $i \in I \backslash i_0$.

Note that mixed strategies and behavior strategies are different. In behavior strategies, roughly speaking, a player 'mixes' the actions at his information sets independently. In contrast, mixed strategies are more general—the probabilities of actions can be conditioned on a player's choices at his other information sets. By Kuhn's (1953) well-known theorem, one can restrict attention to behavior strategies in extensive game with perfect recall—for every mixed strategy combination σ, there exists a behavior strategy combination b that constitutes (via (3.5)) the same distribution over terminal nodes as σ (via (3.3)).

In order to handle extensive games within the framework of strategic games (see Subsection 2.3.2), Selten (1975) introduces agent normal forms of extensive games.

Definition 3.2.3 (Agent normal forms of extensive games).
Let $\Gamma = (T, \lhd, I, P, H, A, p, u)$ be an extensive game. The finite strategic game $G(\Gamma) = (H \backslash H_{i_0}, (A_h)_{h \in H \backslash H_{i_0}}, (u_h)_{h \in H \backslash H_{i_0}})$ with $u_h = u_i$ for all $i \in I \backslash i_0$ and $h \in H_i$ is called the agent normal form of Γ.

In $G(\Gamma)$, $\mathbf{A}$ is the set of pure-strategy combinations, A_h (B_h) is the set of player h's pure (mixed) strategies, and B is the set of mixed strategy combinations. I.e., each player is dissected into agents who control his behavior at the single information sets. This way, of course, some of the structure of Γ, e.g., the sequence of moves, is lost (see Subsection 4.3.4).

3.2.4 Subgames

Let $\Gamma = (T, \lhd, I, P, H, A, p, u)$ be an extensive game and let x be one of its decision nodes. Further, let $T^x \subseteq T$ be the set containing all of x's successors and x itself. The node x induces a subgame of Γ with node set T^x if

$$h(x') \subseteq T^x \tag{3.7}$$

for all $x' \in T^x \backslash Z$. I.e., whenever a player has to move in Γ, he knows whether he is in the subgame or not. The subgame induced by $x \in X$ is denoted $\Gamma^x = (T^x, \lhd^x, I^x, P^x, H^x, A^x, p^x, u^x) \in \mathcal{G}^e$ where its constituents are defined as follows:

The precedence relation $\lhd^x$ is just the restriction of $\lhd$ to T^x. This makes x the root of $(T^x, \lhd^x)$, $Z^x := T^x \cap Z$ its set of terminal nodes, and $X^x := T^x \backslash Z = T^x \cap X$ its set of decision nodes.

The player set I^x contains the player i_0 and all personal players i for which $P_i^x := P_i \cap T^x$ is not empty.

The player partition P^x is the intersection of P with T^x—$P^x := \{P_i^x | i \in I^x\}$.

The information partition H^x is the subset of H that contains the information sets comprising the decision nodes of T^x—$H \supseteq H^x := \{h(x')|x' \in X^x\}$. By (3.7), H^x is a subpartition of P^x. We set $H^x \supseteq H_i^x := \{h(x')|x' \in P_i^x\}$ for all $i \in I^x$.

The action partition A^x is the subset of A that contains the actions of the information sets in H^x—$A^x := \bigcup_{h \in H^x} A_h$. Clearly, A^x is a partition of $T^x \backslash x$. We set $A_i^x := \bigcup_{h \in H_i^x} A_h$ for all $i \in I^x$.

The chance mechanism p^x is the subset of p that contains the probability distributions p_h for all $h \in H_{i_0}^x$.

For all $i \in I^x \backslash i_0^x$, u_i^x is the restriction of u_i to Z^x.

Example 3.2.8.
In $\bar{\Gamma}$ (see Figure 3.1), the node x_4 constitutes a subgame of $\bar{\Gamma}$ as $h(x_4) = \{x_4\} \subseteq \{x_4, z_5, z_6\} = \bar{T}^{x_4}$. In contrast, x_3 does not constitute a subgame—$h(x_3) = \{x_2, x_3\} \not\subseteq \{x_3, z_3, z_4\} = \bar{T}^{x_3}$.

3.2.5 Solution Concepts

In analogy to solution concepts for strategic games (Definition 2.3.4, p. 15), solution concepts for extensive games are defined. Since our focus is on games with perfect recall, solution concepts refer to behavior strategies (see Subsection 3.2.3).

Definition 3.2.4 (Solution concepts for extensive games).
A solution concept L for $\mathcal{G}^e$ assigns a set of behavior strategy combinations $L(\Gamma) \subseteq B$—the solutions—to any extensive game $\Gamma \in \mathcal{G}^e$.

In the following, the equilibrium concepts used in order the assess the weak isomorphisms (to be introduced) as indicated in the Section 3.1 are presented. The first definition extends the Nash equilibrium concept (see Definition 2.3.5, p. 15) to extensive games.

Definition 3.2.5 (Nash equilibria).

Let Γ be an extensive game. A behavior strategy combination $b^* \in B$ is a (Nash) equilibrium if $u_i(b^*) = \max_{b_i \in B_i} u_i(b_i, b^*_{I \setminus \{i_0, i\}})$ for all $i \in I \setminus i_0$.

In many cases, Nash equilibria do not meet our intuitions about sequential rationality. In order to remedy this shortcoming, Selten (1975) introduces the concept of subgame perfect equilibria.

Definition 3.2.6 (Subgame perfect equilibria).

Let Γ be an extensive game. A behavior strategy combination $b^* \in B$ is a subgame perfect equilibrium if it constitutes an equilibrium on every subgame of Γ.

Many extensive games lack non-trivial subgames. Therefore, the concept of subgame perfect equilibria does not refine the Nash equilibrium concept for these games. Selten's (1975) perfect equilibria and Kreps & Wilson's (1982) sequential equilibria fill that gap. Perfect equilibria of extensive games can be characterized within the framework of strategic games: Perfect equilibria of extensive games coincide with the perfect equilibria of their agent normal forms (Selten 1975, Theorem 4).

Definition 3.2.7 (Sequential equilibria).

Let Γ be an extensive game. A system of beliefs is a mapping $\mu : X \to [0,1]$ satisfying $\sum_{x \in h} \mu(x) = 1$ for all $h \in H \setminus H_{i_0}$. A system of beliefs together with a behavior strategy combination is called an assessment. The payoff of a personal player $i(h)$ at $h \in H \setminus H_{i_0}$ with respect to the system of beliefs μ is defined as

$$u_h(\mu, b) := \sum_{x \in h} \mu(x) u_x(b) \tag{3.8}$$

where

$$u_x(b) := \sum_{z \in Z(x)} \frac{\mathrm{prob}(z|b,x)}{\displaystyle\sum_{z' \in Z(x)} \mathrm{prob}(z'|b,x)} u_{i(x)}(z) \tag{3.9}$$

$$\mathrm{prob}(z|b,x) := \begin{cases} \displaystyle\prod_{\substack{a \in a(\psi(z) \setminus \psi(x)) \\ a \notin A_{i_0}}} b_{V(a)}(a) \prod_{\substack{a \in a(\psi(z) \setminus \psi(x)) \\ a \in A_{i_0}}} p_{V(a)}(a) & : z \in Z(x) \\[4mm] 0 & : z \notin Z(x) \end{cases} \tag{3.10}$$

and $Z(x)$ denotes the set of all terminal nodes succeeding x. Note that $V(a)$ is the information set a belongs to. An assessment (μ, b) is said to be sequentially rational if $u_h(\mu, b) \geq u_h(\mu, b'_i b_{-i})$ for all $i \in I \setminus i_0$, $h \in H_i$, and $b'_i \in B_i$.

For all $b \in B^0$, let $\mu(b)$ denote the system of beliefs that is associated with b via Bayes' rule, i.e., we have

$$\mu(b)(x) := \frac{\text{prob}(x|b)}{\text{prob}(h(x)|b)} = \frac{\displaystyle\sum_{\mathbf{a}^+ \in \mathbf{A}^+(x)} \text{prob}(\mathbf{a}^+|b)}{\displaystyle\sum_{\mathbf{a}^+ \in \mathbf{A}^+(h(x))} \text{prob}(\mathbf{a}^+|b)} \qquad (3.11)$$

where $\mathbf{A}^+(x)$ and $\mathbf{A}^+(h)$ denote the set of extended action combinations that lead to terminal nodes succeeding x and h, respectively.

An assessment (μ, b) is said to be consistent if there is some sequence $(b^n)_{n \in \mathbb{N}}$ such that $b^n \in B^0$ and $\lim_{n \to \infty}(\mu(b^n), b^n) = (\mu, b)$.

An assessment (μ, b) is a sequential equilibrium if it is consistent and sequentially rational.

3.3 Invariance with Respect to Weak Isomorphisms

3.3.1 Weak Isomorphisms of Extensive Games

In order to identify corresponding actions of the two players in his evolutionary stability concept for two-player extensive games, Selten (1983) introduces symmetries for this class of games that are based on bijective mappings of the action partition onto itself. Oh (1995) extends Selten's definition to a general notion of symmetries of extensive games. Casajus' (1998) definition departs from Oh in making some corrections and amendments that are in line with Selten's original definition.

Definition. The following definition extends Casajus' definition of symmetries to weak isomorphisms of extensive games in a straightforward manner.

Definition 3.3.1 (Weak isomorphisms of extensive games).
Let $G = (T, \lhd, I, P, H, A, p, u)$ and $\bar{\Gamma} = (\bar{T}, \bar{\lhd}, \bar{I}, \bar{P}, \bar{H}, \bar{A}, \bar{p}, \bar{u})$ be extensive games. A weak isomorphism from Γ onto $\bar{\Gamma}$ is a bijection $r : A \to \bar{A}$ with the following properties: There are bijections $\nu : H \to \bar{H}$, $\pi : I \to \bar{I}$, and $\theta : Z \to \bar{Z}$ such that

ISA $r(A_h) = \bar{A}_{\nu(h)}$ for all $h \in H$,

PL $r(A_i) = \bar{A}_{\pi(i)}$ for all $i \in I$,

CPL $r(A_{i_0}) = \bar{A}_{i_0}$,

CPR $p_h(a) = \bar{p}_{\nu(h)}(r(a))$ for all $h \in H_{i_0}$ and $a \in A_h$,

PTH $r(a(\psi(z))) = \bar{a}(\bar{\psi}(\theta(z)))$ for all $z \in Z$,

PY   there are constants $\alpha_i, \beta_i \in \mathbb{R}$, $\alpha_i > 0$ for all $i \in I \setminus i_0$ such that

$$\bar{u}_{\pi(i)}(\theta(z)) = \alpha_i u_i(z) + \beta_i$$

for all $z \in Z$ and $i \in I \setminus i_0$.

Since the identity mapping is an automorphism of any extensive game and since all mappings involved are bijective, weak isomorphisms constitute an equivalence relation on $\mathcal{G}^e$. Two extensive games Γ and $\bar{\Gamma}$ are said to be weakly isomorphic if there is a weak isomorphism from Γ onto $\bar{\Gamma}$.

In the following, the conditions of Definition 3.3.1 are discussed in detail.

Conditions ISA, PL, CPL. Condition **ISA** (short for Information Set-Action structure) is equivalent to requiring that actions in Γ should be mapped onto actions of the same information set in $\bar{\Gamma}$ if and only if they belong to the same information set in Γ. Thereby, a unique bijection $\nu : H \to \bar{H}$ is determined. This property corresponds to Selten's condition (d); it is implied by Peleg et al. (1999) and Elmes & Reny (1994). In addition, any bijection $r : A \to \bar{A}$ satisfying **ISA** induces a unique bijection $\mathbf{r}^+ = (\mathbf{r}_{\bar{h}}^+)_{\bar{h} \in \bar{H}}$, $\mathbf{r}^+ : \mathbf{A}^+ \to \bar{\mathbf{A}}^+$ with

$$\mathbf{r}_{\nu(h)}^+(\mathbf{a}^+) := r(\mathbf{a}_h^+) \tag{3.12}$$

for all $\mathbf{a}^+ \in \mathbf{A}^+$ and $h \in H$ where ν is determined by r according to **ISA**.

Condition **PL** (short for **PL**ayer structure) is equivalent to requiring that actions in Γ should be mapped onto actions of the same player in $\bar{\Gamma}$ if and only if they belong to same player in Γ. Thereby, a unique bijection $\pi : I \to \bar{I}$ is determined. This property corresponds to Selten's condition (b); it is also implied by Peleg et al. (1999) and Elmes & Reny (1994). Together with condition **ISA** this implies

$$\nu(H_i) = \bar{H}_{\pi(i)} \tag{3.13}$$

for all players i—ν maps all information sets of player i in Γ onto the information sets of player $\pi(i)$ in $\bar{\Gamma}$.

Condition **CPL** (short for Chance **PL**ayer structure) requires that actions of the chance mechanism in Γ and only these actions should be mapped to actions of the $\bar{\Gamma}$'s chance mechanism, or the other way round, actions of personal players in Γ and only these actions should be mapped to actions of $\bar{\Gamma}$'s personal players. In view of condition **PL**, condition **CPL** implies $\pi(i_0) = \bar{i}_0$ for the bijection π determined by **PL**. This property corresponds to the first part of Selten's condition (a). In addition, any bijection $r : A \to \bar{A}$ satisfying **ISA** and **CPL** induces a unique bijection $\mathbf{r} = (\mathbf{r}_{\bar{h}})_{\bar{h} \in \bar{H} \backslash \bar{H}_{i_0}}$, $\mathbf{r} : \mathbf{A} \to \bar{\mathbf{A}}$ with

$$\mathbf{r}_{\nu(h)}(\mathbf{a}) := r(\mathbf{a}_h) \tag{3.14}$$

for all $h \in H \backslash H_{i_0}$ and $\mathbf{a} \in \mathbf{A}$ which is extended to B by setting

$$\mathbf{r}_{\nu(h)}(b)(r(a)) := b_h(a) \tag{3.15}$$

for all $h \in H \backslash H_{i_0}$ and $a \in A_h$.

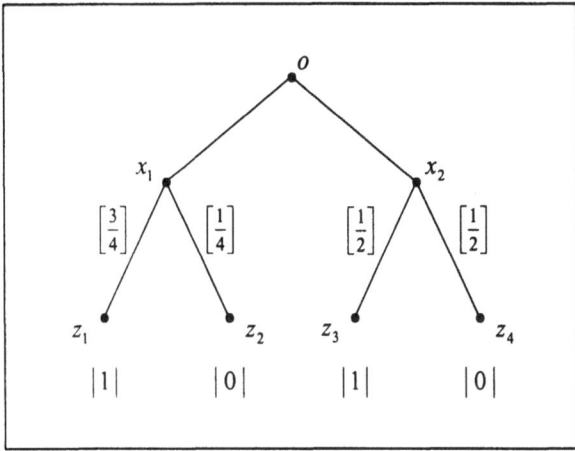

Fig. 3.4. Condition **CPR** is indispensable

Condition CPR. Condition **CPR** (short for Chance moves' **PR**obabilities) requires that actions of the chance mechanism in Γ are mapped onto actions of $\bar{\Gamma}$'s chance mechanism only if they get the same probabilities. Clearly, this only makes sense together with condition **ISA** and **CPL**. Again, this condition is implied by Peleg et al. (1999), and it corresponds to the second part of Selten's condition (a).

We take the view that the conditions **ISA**, **PL**, and **CPL** are indispensable for any kind of isomorphism that preserves the *structure* of extensive games. As the following example shows, in addition, we hardly can do without **CPR**. Note also the role of **CPR** in the proof of Lemma 3.3.2 (p. 71).

Example 3.3.1.
Consider the game in Figure 3.4 where player 1 controls node o; and the nodes x_1 and x_2 are controlled by the chance mechanism according to the probabilities in brackets. Let $r : A \to A$ be such that $r(\{x_1\}) = \{x_2\}, r(\{z_1\}) = \{z_3\}, r(\{z_2\}) = \{z_4\}$ and vice versa. While r violates **CPR**, it satisfies **PTH, ISA, PL, CPL**, and **PY**. But taking $\{x_1\}$ gives player 1 an expected payoff of $\frac{3}{4}$ and taking $\{x_2\}$ gives $\frac{1}{2}$. Therefore, action $\{x_1\}$ should not be mapped onto $\{x_2\}$ by an isomorphism.

Condition PTH. Condition **PTH** (short for **Pa**T**H** structure) requires that plays expressed by actions in Γ should be mapped onto plays expressed by actions in $\bar{\Gamma}$. Thereby, a unique bijection $\theta : Z \to \bar{Z}$ would be induced. This condition corresponds to Selten's condition (e). He only requires the existence of a terminal node $\bar{z} \in \bar{Z}$ for each terminal node $z \in Z$ such that $r(a(\psi(z))) = \bar{a}(\bar{\psi}(\bar{z}))$. In view of Remark 3.2.1, this $\bar{z}$ is unique. Hence, Selten's condition and **PTH** are equivalent.

PTH and **ISA** together establish the following property of the mapping r^+.

Lemma 3.3.1.
*Let Γ and $\bar{\Gamma}$ be extensive games; and let r be a bijection $r : A \to \bar{A}$ that satisfies **ISA** and **PTH**. Then,*

PTH$^-$ $\bar{z}(r^+(\mathbf{a}^+)) = \theta(z(\mathbf{a}^+))$ *holds for all* $\mathbf{a}^+ \in \mathbf{A}^+$

*where r^+ is given by r via (3.12), and θ is determined by **PTH**.*

Proof. For all $\mathbf{a}^+ \in \mathbf{A}^+$, we have

$$\begin{aligned}
\{r_{\bar{h}}^+(\mathbf{a}^+)|\bar{h} \in \bar{H}\} &= \{r_{\nu(h)}^+(\mathbf{a}^+)|h \in H\} && \nu \text{ is bijective,} \\
&= \{r(\mathbf{a}_h^+), h \in H\} && (3.12), \\
&= r\left(\{\mathbf{a}_h^+, h \in H\}\right) \\
&\supseteq r(a(\psi(z(\mathbf{a}^+)))) && (3.1), \\
&\supseteq \bar{a}(\bar{\psi}(\theta(z(\mathbf{a}^+)))) && \textbf{PTH},
\end{aligned}$$

and therefore $\bar{z}(r^+(\mathbf{a}^+)) = \theta(z(\mathbf{a}^+))$ by (3.1) and (3.12). $\square$

By condition **PTH**, any bijection $r : A \to \bar{A}$ induces a unique bijection θ between the sets of terminal nodes. Lemma 3.3.1 points into a direction in which condition **PTH** can be weakened without loosing this property. Consider the following condition instead: Extended pure-strategy combinations that lead to the same result (terminal node) in Γ and only these should be mapped to extended pure-strategy combinations in $\bar{\Gamma}$ that lead to the same result (terminal node). This requirement is equivalent to the existence of a unique bijection $\theta : Z \to \bar{Z}$ such that

PTH$^-$ $\theta(z(\mathbf{a}^+)) = \bar{z}(r^+(\mathbf{a}^+))$ holds for all $\mathbf{a}^+ \in \mathbf{A}^+$

where r^+ is determined by r and **ISA** via (3.12).

In a sense, **PTH$^-$** seems to be the weakest possible requirement that preserves the relation between actions and outcomes (terminal nodes). In view of Lemma 3.3.1, **PTH$^-$** is implied by **PTH** and **ISA**, but **ISA** and **PTH$^-$** together do not imply **PTH**. A counterexample:

Example 3.3.2.
Consider the game in Figure 3.5 where player 1 controls all decision nodes. Let $r : A \to A$ be such that $r(\{x_1\}) = \{x_1\}$, $r(\{x_2\}) = \{x_2\}$, $r(\{z_1\}) = \{z_2\}$, $r(\{z_2\}) = \{z_1\}$. Obviously, r satisfies **PTH$^-$** (as well as **ISA**, **PL**, **CPL**, **CPR**, and **PY**), and we have $\theta(z_1) = z_1$ and $\theta(z_2) = z_2$. But $r(a(\psi(z_1))) = \{\{x_1\}, \{z_2\}\} \neq \{\{x_1\}, \{z_1\}\} = a(\psi(\theta(z_1)))$, i.e., **PTH** does not hold.

In addition, the mapping r changes the preference order of the terminal nodes reached by the immediately preceding actions: The action $\{z_2\}$ leads to the terminal node z_2 which is preferred to terminal node z_1 reached by action $\{z_1\}$. But the terminal node z_1 reached by action $r(\{z_2\}) = \{z_1\}$ is preferred to z_2 reached by action $r(\{z_1\}) = \{z_2\}$. Therefore, one could argue

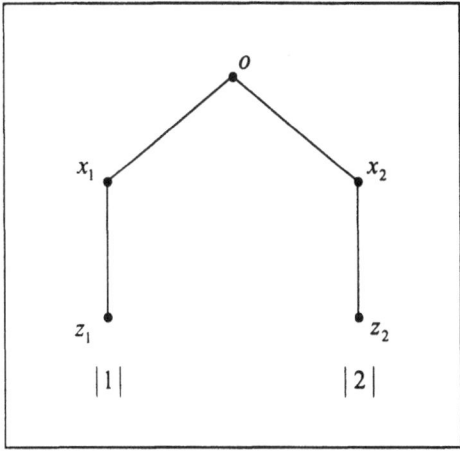

Fig. 3.5. Condition **PTH⁻** does not imply **PTH**

that **PTH⁻** is to weak a requirement. But the counterexample involves a pathological extensive game (see Definition 3.2.1, p. 60). Without this peculiarity, it turns out that **PTH⁻** and **PTH** coincide in view of **ISA—PTH⁻** does not weaken **PTH** substantially.

Theorem 3.3.1.
*Let Γ and $\bar{\Gamma}$ be non-pathological extensive games; and let r be a bijection r : $A \to \bar{A}$ that satisfies **ISA** and **PTH⁻**. Then, r also satisfies **PTH**.*

Proof. Let r be a bijection $r : A \to \bar{A}$ that satisfies **ISA** and **PTH⁻**. By definition, **PTH⁻** induces a bijection $\theta : Z \to \bar{Z}$ such that $\theta(z(\mathbf{a}^+))$ $= \bar{z}(\mathbf{r}^+(\mathbf{a}^+))$ for all $\mathbf{a}^+ \in \mathbf{A}^+$ where $\mathbf{r}^+$ is determined by r and **ISA** via (3.12). Since $z(\mathbf{A}^+) = Z$, it is sufficient to show that we have $r(a(\psi(z(\mathbf{a}^+))))$ $= \bar{a}(\bar{\psi}(\bar{z}(\mathbf{r}^+(\mathbf{a}^+))))$ for all $\mathbf{a}^+ \in \mathbf{A}^+$. Suppose on the contrary there were some $\mathbf{a}^+ \in \mathbf{A}^+$ and $h \in H$ such that $\mathbf{a}_h^+ \in r(a(\psi(z(\mathbf{a}^+))))$ but $r(\mathbf{a}_h^+) \notin \bar{a}(\bar{\psi}(\bar{z}(\mathbf{r}^+(\mathbf{a}^+))))$. Since Γ is non-pathological, some $\mathbf{a}^{+'} \in \mathbf{A}^+$ existed such that $\mathbf{a}_{h'}^+ = \mathbf{a}_{h'}^{+'}$ for all $h' \in H \backslash h$ and $\mathbf{a}_h^+ \neq \mathbf{a}_h^{+'}$. By **ISA** and (3.1), we have $r(\mathbf{a}_h^{+'}) \notin \bar{a}(\bar{\psi}(\bar{z}(\mathbf{r}^+(\mathbf{a}^+))))$ and therefore $\bar{z}(\mathbf{r}^+(\mathbf{a}^+)) = \bar{z}(\mathbf{r}^+(\mathbf{a}^{+'}))$. By (3.1), on the other hand, we have $z(\mathbf{a}^+) \neq z(\mathbf{a}^{+'})$. But this contradicts the bijectivity of θ. Thus, $r(a(\psi(z(\mathbf{a}^+)))) \subseteq \bar{a}(\bar{\psi}(\bar{z}(\mathbf{r}^+(\mathbf{a}^+))))$. Analogously, one can show that $r(a(\psi(z(\mathbf{a}^+)))) \supseteq \bar{a}(\bar{\psi}(\bar{z}(\mathbf{r}^+(\mathbf{a}^+))))$. Therefore, $r(a(\psi(z(\mathbf{a}^+))))$ $= \bar{a}(\bar{\psi}(\bar{z}(\mathbf{r}^+(\mathbf{a}^+)))) = \bar{a}(\bar{\psi}(\theta(z(\mathbf{a}^+))))$ for all $\mathbf{a}^+ \in \mathbf{A}^+$, i.e., **PTH** holds. □

Condition PY. In view of condition **PL** and **PTH**, condition **PY** (short for **Pa**Yoff structure) requires that the preferences of corresponding players (connected by π) over corresponding results/terminal nodes (connected by θ) should be the same. I.e., the payoff functions can obtained from each other by positive affine transformations. This property corresponds to Selten (1983) condition (f). As Peleg et al. (1999) and Elmes & Reny (1994), and in contrast to Oh (1995), he fixes all α_i at 1 and all β_i at 0. The next two lemmas show that one can replace condition **PY** by a condition that is similar to (2.5).

Lemma 3.3.2.
Let r be a weak isomorphism from Γ onto $\bar{\Gamma}$. For all $\mathbf{a}^+ \in \mathbf{A}^+$ and $b \in B$, we have $\mathrm{prob}(\mathbf{a}^+|b) = \mathrm{prob}(\mathbf{r}^+(\mathbf{a}^+)|\mathbf{r}(b))$ where $\mathbf{r}$ and $\mathbf{r}^+$ are given by r via (3.14) and (3.12) respectively.

Proof. For all $\mathbf{a}^+ \in \mathbf{A}^+$ and $b \in B$ we have

$$
\begin{aligned}
\mathrm{prob}(\mathbf{r}^+(\mathbf{a}^+)|\mathbf{r}(b)) &= \prod_{\bar{h} \in \bar{H}_{i_0}} \bar{p}_{\bar{h}}(\mathbf{r}_{\bar{h}}^+(\mathbf{a}^+)) \prod_{\bar{h} \in \bar{H} \backslash \bar{H}_{i_0}} \mathbf{r}_{\bar{h}}(b)(\mathbf{r}_{\bar{h}}^+(\mathbf{a}^+)) \\
&= \prod_{h \in H_{i_0}} \bar{p}_{\nu(h)}(\mathbf{r}_{\nu(h)}^+(\mathbf{a}^+)) \prod_{h \in H \backslash H_{i_0}} \mathbf{r}_{\nu(h)}(b)(\mathbf{r}_{\nu(h)}^+(\mathbf{a}^+)) \\
&= \prod_{h \in H_{i_0}} \bar{p}_{\nu(h)}(r(\mathbf{a}_h^+)) \prod_{h \in H \backslash H_{i_0}} \mathbf{r}_{\nu(h)}(b)(r(\mathbf{a}_h^+)) \\
&= \prod_{h \in H_{i_0}} p_h(\mathbf{a}_h^+) \prod_{h \in H \backslash H_{i_0}} b_h(\mathbf{a}_h^+) \\
&= \mathrm{prob}(\mathbf{a}^+|b)
\end{aligned}
$$

where the equations follow from (3.4), (3.14), and (3.12); ν's bijectivity and **CPL**; (3.14) and (3.12); **CPR** and (3.15); and (3.4) respectively. $\square$

Lemma 3.3.3.
*Within Definition 3.3.1, condition **PY***

1. *is equivalent to $\bar{u}_{\pi(i)}(\mathbf{r}^+(\mathbf{a}^+)) = \alpha_i u_i(\mathbf{a}^+) + \beta_i$ holding for all $i \in I \backslash i_0$ and $\mathbf{a}^+ \in \mathbf{A}^+$,*
2. *is equivalent to $\bar{u}_{\pi(i)}(\mathbf{r}(\mathbf{a})) = \alpha_i u_i(\mathbf{a}) + \beta_i$ holding for all $\mathbf{a} \in \mathbf{A}$ and $i \in I \backslash i_0$ in the case that $p = \emptyset$,*
3. *implies $\bar{u}_{\pi(i)}(\mathbf{r}(\mathbf{a})) = \alpha_i u_i(\mathbf{a}) + \beta_i$ holding for all $\mathbf{a} \in \mathbf{A}$ and $i \in I \backslash i_0$.*

Proof. Let r be a weak isomorphism from Γ onto $\bar{\Gamma}$. Since we have $z(\mathbf{A}^+) = Z$ and $\overset{-1}{z}(Z) = \mathbf{A}^+$, Lemma 3.3.1 implies the first claim. Therefore, the second claim also holds as $p = \emptyset$ implies $\mathbf{A} = \mathbf{A}^+$ and $\mathbf{r} = \mathbf{r}^+$. For all $b \in B$ and $i \in I \backslash i_0$ we have

$$
\begin{aligned}
\bar{u}_{\pi(i)}(\mathbf{r}(b)) &= \sum_{\bar{\mathbf{a}}^+ \in \bar{\mathbf{A}}^+} \mathrm{prob}(\bar{\mathbf{a}}^+|\mathbf{r}(b)) \cdot \bar{u}_{\pi(i)}(\bar{\mathbf{a}}^+) \\
&= \sum_{\mathbf{a}^+ \in \mathbf{A}^+} \mathrm{prob}(\mathbf{r}^+(\mathbf{a}^+)|\mathbf{r}(b)) \cdot \bar{u}_{\pi(i)}(\mathbf{r}(\mathbf{a}^+)) \\
&= \sum_{\mathbf{a}^+ \in \mathbf{A}^+} \mathrm{prob}(\mathbf{a}^+|b) \cdot \bar{u}_{\pi(i)}(\mathbf{r}(\mathbf{a}^+))
\end{aligned}
$$

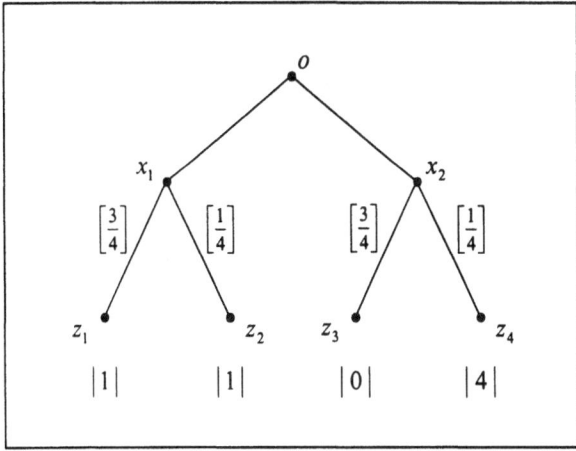

Fig. 3.6. PY cannot be defined in terms of behavior strategies

$$= \alpha_i \left(\sum_{\mathbf{a}^+ \in \mathbf{A}^+} \text{prob}(\mathbf{a}^+ | b) \cdot u_i(\mathbf{a}^+) \right) + \beta_i$$
$$= \alpha_i u_i(b) + \beta_i$$

where the equations follow from (3.6), $\mathbf{r}^+$'s bijectivity, Lemma 3.3.2, the first claim, and (3.6), respectively. This implies the third claim. As the following example shows, the converse does not hold in general. $\square$

Example 3.3.3.
Consider the game in Figure 3.6 where player 1 controls all decision nodes. Let $r : A \to A$ be such that $r(\{x_1\}) = \{x_2\}$, $r(\{z_1\}) = \{z_3\}$, $r(\{z_2\}) = \{z_4\}$ and vice versa. Since $u_1((\{x_1\})) = u_1((\{x_2\})) = 1$, r meets the third claim (as well as **PTH, ISA, PL, CPL**, and **CPR**); but it violates **PY**—while $u_1(z_1) = u_1(z_2)$, we have $u_1(\theta(z_1)) \neq u_1(\theta(z_2))$.

3.3.2 Weak Isomorphism Invariance

The following definition ensures that the solutions of extensive games are independent of the game theorist's labelling and the game theorist's representation of the players' preferences (remember Subsection 2.3.4).

Definition 3.3.2 (Invariance with respect to weak isomorphisms).
A solution concept L for $\mathcal{G}^e$ is invariant with respect to weak isomorphisms if, for every weak isomorphism r from $\Gamma \in \mathcal{G}^e$ onto $\bar{\Gamma} \in \mathcal{G}^e$, we have $\mathbf{r}(L(\Gamma)) = L(\bar{\Gamma})$ for $\mathbf{r}$ given by r via (3.14).

3.3.3 Weak Symmetry Invariance

Weak automorphisms of extensive games are called weak symmetries. The definitions below are very similar to their respective counterparts for strategic games (see Subsection 2.3.5).

Definition 3.3.3 (Weak symmetry invariance).
A behavior-strategy combination b of an extensive game Γ is called weakly symmetry invariant if $\mathbf{r}(b) = b$ for all mappings $\mathbf{r}$ given by weak symmetries r of Γ via (3.15).

As in strategic games, invariance with respect to weak isomorphisms implies that a one-point solution concept assigns a weakly symmetry invariant behavior strategy combination to any extensive game.

Definition 3.3.4 (Weakly symmetric actions).
Two actions $a, a' \in A$ of Γ are weakly symmetric, if there is a symmetry r of Γ such that $r(a) = a'$.

In view of (3.15), weak symmetry invariance can be characterized as follows.

Corollary 3.3.1.
A behavior-strategy combination is weakly symmetry invariant if and only if weakly symmetric strategies get the same probabilities.

The next theorem and corollary establish the relation between weak symmetry in extensive games and symmetry in their agent normal form (Definition 3.2.3, p. 64).

Theorem 3.3.2.
Weakly symmetric actions in Γ are also symmetric in Γ's agent normal form representation $G(\Gamma)$.

Proof. Let r be a weak symmetry of Γ, let π, ν, and θ be the bijections induced by r according to Definition 3.3.1, and let $(H\backslash H_{i_0}, (A_h)_{h \in H \backslash H_{i_0}}, (u_h)_{h \in H \backslash H_{i_0}})$ be the agent normal form of Γ. Consider the restriction ν' of ν to $H\backslash H_0$. By **PL** and (3.13), ν' is a permutation of $H\backslash H_{i_0}$. By **ISA**, r can be split into bijections $r_h : A_h \to A_{\nu'(h)}$ with $r_h(a) := r(a)$ for all $h \in H\backslash H_0$ and $a \in A$. By Lemma 3.3.3, we have $u_{\nu'(h)}(\mathbf{r}(\mathbf{a})) = \alpha_i u_h(\mathbf{a}) + \beta_i$ for all $h \in H\backslash H_0$ and $\mathbf{a} \in \mathbf{A}$. Since by (3.14) $\mathbf{r}$ is induced by the mappings r_h in accordance with (2.6), $(\nu', (r_h)_{h \in H\backslash H_0})$ is a symmetry of G. In view of Definitions 2.3.9 and 3.3.4, this proves the claim. $\square$

Since agent normal form representations ignore essential parts of the structure of extensive forms, the converse does not hold in general (see Example 3.3.3). A much weaker result in the opposite direction is given by Theorem 3.5.2.

By Theorem 3.3.2, the weak symmetry of actions is transferred from an extensive game to its agent normal form. The Corollaries 2.3.1 and 3.3.1 then imply that symmetry invariance transfers in the opposite direction.

Corollary 3.3.2.
If a behavior-strategy combination is symmetry invariant in the agent normal form of an extensive game, then it is weakly symmetry invariant in the extensive game.

Now, we can establish the following strong existence result.

Theorem 3.3.3.
Every extensive game has a weakly symmetry invariant perfect equilibrium.

Proof. By Corollary 2.3.3 (p. 24), the agent normal form $G(\Gamma)$ of every extensive game Γ has a symmetry invariant perfect equilibrium b; by Selten (1975, Theorem 4), b is also a perfect equilibrium of Γ, and by Corollary 3.3.2, b is weakly symmetry invariant. $\square$

The existence of weakly symmetry invariant (a) equilibria, (b) sequential equilibria, (c) subgame perfect equilibria directly follows from Selten (1975, Lemma 3), Kreps & Wilson (1982, Proposition 5), and Selten (1975, Theorem 2), respectively.

3.3.4 The Order of Actions

As already mentioned, weak isomorphisms do not preserve the order of moves in a strict sense. First, the notion of strict order is made explicit. The following subsections explore some properties of weak isomorphisms that are related to the order of actions.

Respecting the Strict Order. As the example below shows, weak isomorphisms do not always respect the strict order of actions, i.e., in the sense that the following strengthening of condition **PTH** is satisfied: Ordered plays expressed by actions in Γ should be mapped onto ordered plays expressed by actions in $\bar{\Gamma}$. I.e., the order of actions within a play is not affected by an isomorphism. Note that Peleg et al.'s (1999) strong isomorphisms meet this requirement (see Section 3.4). Thereby, again, a unique bijection $\theta : Z \to \bar{Z}$ would be induced such that

PTH$^+$ $r(a(\psi(z))) = \bar{a}(\bar{\psi}(\theta(z)))$ for all $z \in Z$

where r is extended to ordered plays expressed by actions such that

$$r(a(\psi(z))) := (r(a(\psi_k(z))))_{k \in \{1,\dots,|\psi(z)|\}} \tag{3.16}$$

holds for all $z \in Z$. Clearly, **PTH$^+$** implies **PTH**. But the converse does not hold. A counterexample:

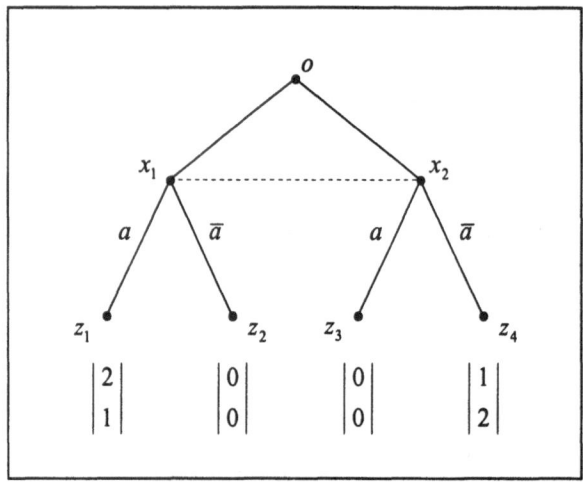

Fig. 3.7. Traditional extensive representation of the Battle-of-the-Sexes game

Example 3.3.4.
Consider the traditional extensive representation of the Battle-of-the Sexes game in Figure 3.7, where player 1 controls the information set $\{o\}$, and player 2 controls the information set $\{x_1, x_2\}$ with the actions $a = \{z_1, z_3\}$ and $\bar{a} = \{z_2, z_4\}$. Let $r : A \to A$ be such that $r(\{x_1\}) = \{z_2, z_4\}$, $r(\{x_2\}) = \{z_1, z_3\}$ and vice versa. Obviously, r satisfies **PTH** (as well as **ISA**, **PL**, **CPL**, **CPR**, and **PY**), and we have $\theta(z_1) = z_4$. But $r(a(\psi(z_1))) = r((\{x_1\}, \{z_1, z_3\})) = (\{z_2, z_4\}, \{x_2\}) \neq (\{x_2\}, \{z_2, z_4\}) = a(\psi(z_4))$, i.e., **PTH$^+$** does not hold.

Subgame Preservation. Although weak isomorphisms do not preserve the order of moves in general, one should require that they preserve the subgame structure—exactly the actions of one subgame should be mapped again onto the actions of one subgame. Thereby, the hierarchy of subgames also would be preserved. As the following example shows, this property does not hold in general for weak isomorphisms.

Example 3.3.5.
Reconsider the game in Figure 3.5 (p. 71) and its weak symmetry r with $r(\{x_1\}) = \{z_1\}$, $r(\{x_2\}) = \{z_2\}$, and vice versa. The action partition $A^{x_1} = \{\{z_1\}\}$ of subgame Γ^{x_1} is mapped onto $\{\{x_1\}\}$ which is clearly not the action partition of another subgame.

Again, the counterexample involves a pathological game. Adapting Selten's (1983, Theorem 1) proof, one can establish the following theorem.

Theorem 3.3.4.
Let Γ and $\bar{\Gamma}$ be non-pathological extensive games; and let r be a bijection

$r : A \rightarrow \bar{A}$ that satisfies **ISA** and **PTH**. Then, r maps the subgames of Γ onto subgames of $\bar{\Gamma}$ in the following sense: If $A^x \subseteq A$ is the action partition of some subgame Γ^x of Γ, then $r(A^x)$ is the action partition of some subgame of $\bar{\Gamma}$.

While perfect equilibria of extensive games induce perfect equilibria of subgames (Selten 1975, Theorem 2), a similar property does not hold for weak symmetry invariance. A counterexample:

Example 3.3.6.
Consider the game in Figure 3.8 as an extension of the Battle-of-the-Sexes in Figure 3.7 in which the additional decision node $\bar{o}$ is controlled by player 1. The weak symmetry r given by $r(\{x_1\}) = \{x_2, x_4\}$, $r(\{x_2\}) = \{x_1, x_3\}$, $r(\{x_2, x_4\}) = \{x_1\}$, and $r(\{x_1, x_3\}) = \{x_2\}$ establishes the weak symmetry of $\{x_1\}$ and $\{x_2, x_4\}$ in the subgame Γ^o according to Definition 3.3.4. Therefore, coordination at z_1 —the behavior strategy $(\{x_1\}, \{x_1, x_3\})$ is not weakly symmetry invariant in the subgame induced by o.

But in the whole game $(\Gamma^{\bar{o}})$, choosing action $\{o\}$ and coordinating at z_1 is a weakly symmetry invariant: In view of condition **PL**, $\{x_1\}$ and $\{x_2, x_4\}$ cannot be symmetric as player 1 has four actions, whereas player 2 has only two.

This example also shows that it is possible to support some kind focal-point reasoning within the framework of weak symmetries. One could argue that the existence of actions outside of subgame Γ^o makes player 1 the salient one. Of course, this does not necessarily justify that a player should take one rather than another action in the subgame.

Now, one could ask whether the converse holds: Is any weakly symmetry invariant behavior strategy combination of a subgame part of a weakly symmetry invariant behavior strategy combination of the whole game? The following example shows that this is not necessarily so.

Example 3.3.7.
Consider another extension of the Battle-of-the-Sexes game in Figure 3.9 where $\bar{o}$ is controlled by player 1, and $\bar{x}$ is controlled by player 2. As argued in Example 3.3.6, coordination at z_1 is weakly symmetry invariant in $\Gamma^{\bar{x}}$. However, the symmetry r given by $r(\{x_1\}) = \{x_2, x_4\}$, $r(\{x_2\}) = \{x_1, x_3\}$, $r(\{x_2, x_4\}) = \{x_1\}$, $r(\{x_1, x_3\}) = \{x_2\}$, $r(\{\bar{x}\}) = \{o\}$, and $r(\{o\}) = \{\bar{x}\}$, establishes the weak symmetry of $\{x_1\}$ and $\{x_2, x_4\}$ in the whole game $(\Gamma^{\bar{o}})$. Hence, coordination at z_1 cannot be weakly symmetry invariant in $\Gamma^{\bar{o}}$.

Again, the counterexample involves a pathological game. Without this peculiarity, we have:

Theorem 3.3.5.
Let Γ be a non-pathological extensive game and let Γ^x be the subgame of Γ at some decision node x. Then, any weakly symmetry invariant behavior

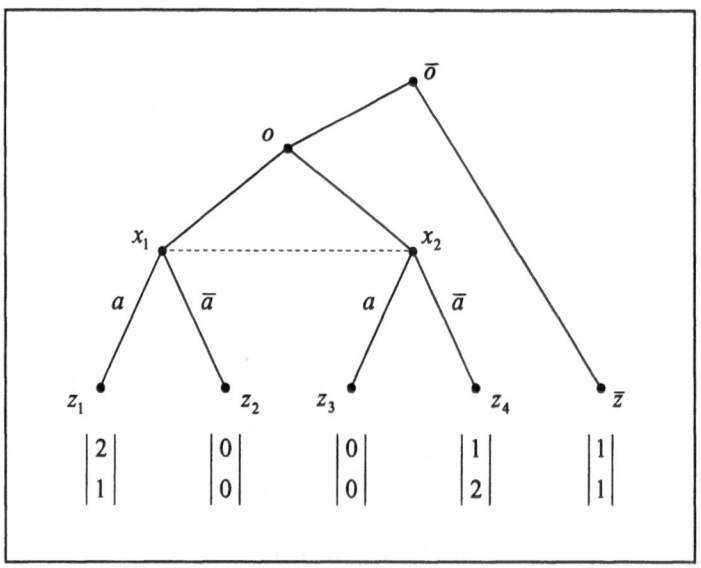

Fig. 3.8. Battle of the Sexes, extended

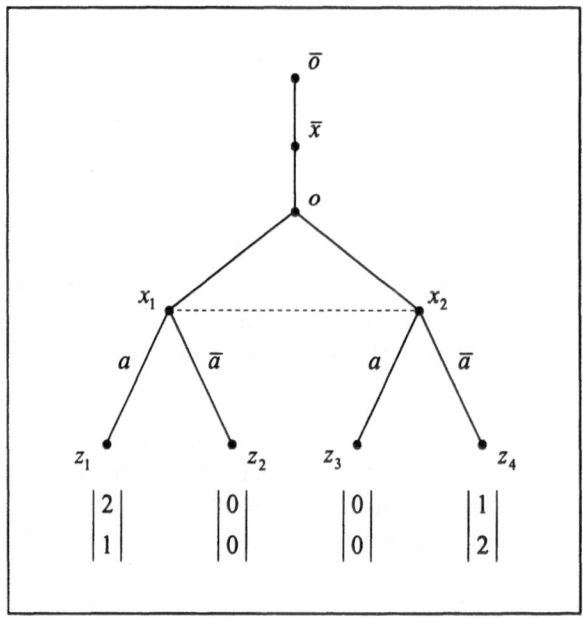

Fig. 3.9. Battle of the Sexes, once more extended

strategy combination of Γ^x is part of a weakly symmetry invariant behavior strategy combination of Γ.

Proof. In view of Corollary 3.3.1 (p. 73), this claim can be expressed as follows: Actions that are weakly symmetric in an extensive game also are weakly symmetric in any subgame containing them.

Let Γ^x be a subgame of Γ for some decision node x, and let a and a' be actions from Γ^x that are weakly symmetric in Γ. Let r be a weak symmetry of Γ that establishes the weak symmetry of a and a' according to Definition 3.3.4, i.e., $r(a) = a'$. By Theorem 3.3.4, in non-pathological cases, A^x is mapped onto another subgame's action set. Since action sets of subgames are either disjoint or one is a subset of the other, by $r(a) = a'$, we have $r(A^x) = A^x$. It remains to show that r's restriction to A^x—r^x—is a weak symmetry of Γ^x. Obviously, r^x is bijective and inherits the properties **ISA**, **CPL**, **CPR** from r. For all $i \in I^x$ we have $r(A_i^x) = r(A_i \cap A^x) = r(A_i) \cap r(A^x) = A_{\pi(i)} \cap A^x = A_{\pi(i)}^x$—$\pi(I^x) = I^x$. Thus, **PL** holds. Since $a(\psi^x(z)) = a(\psi(z)) \cap A^x$ for all $z \in Z^x$, we have $r^x(a(\psi^x(z))) = r(a(\psi(z)) \cap A^x) = r(a(\psi(z))) \cap A^x = a(\psi(\theta(z))) \cap A^x = a(\psi^x(\theta(z)))$. Therefore, **PTH** and **PY** hold. Thus, r^x is a weak symmetry of Γ^x. □

Respecting the Essential Order. As indicated in Section 3.1, we apply the following criterion for assessing whether a certain concept of isomorphisms for extensive games respects the order of moves to the essential extent:

Definition 3.3.5 (Respecting the essential order of moves).

A concept of isomorphisms of extensive games respects the order of moves to the essential extent if the solution concepts under consideration—especially solution concepts that explicitly refer to the sequential nature of moves—are invariant with respect to these isomorphisms (see Definition 3.3.2, p. 73). Besides Nash equilibria and perfect equilibria, we consider subgame perfect equilibria (Selten 1975) and sequential equilibria (Kreps & Wilson 1982).

In view of Lemma 3.3.3 (p. 72), weak isomorphisms always carry (Nash) equilibria of extensive games into (Nash) equilibria. So the Nash equilibrium concept is invariant with respect to weak isomorphisms. As the following example shows, this is not true for subgame perfect equilibria. By Theorem 3.3.4, it should be clear that subgame perfection as a solution concept is invariant with respect to weak isomorphisms in non-pathological cases.

Example 3.3.8.
Consider the game Γ in Figure 3.10, where the information sets $\{o\}$ and $\{x_2, x_3, x_4\}$ are controlled by player 1, and the information set $\{x_1\}$ is controlled by player 2. Since there is no non-trivial subgame, any Nash equilibrium of the game is subgame perfect. So, the behavior strategy combination $b = (\{x_4\}, \{z_1, z_2, z_3\}, \{x_3\})$ is a subgame perfect equilibrium.

Consider now the game $\bar{\Gamma}$ in Figure 3.11 where the information sets $\{o\}$ and $\{x_1\}$ are controlled by player 1, and information set $\{x_2\}$ is controlled

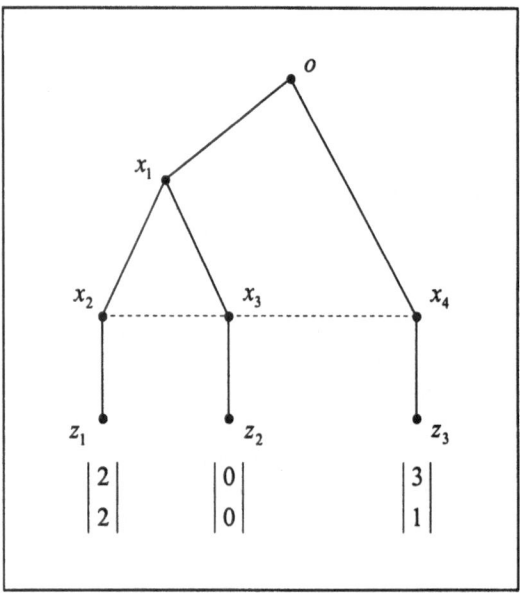

Fig. 3.10. Subgame perfection is not invariant with respect to isomorphisms #1

by player 2. This game has a subgame at x_2. Since $(\{z_1\})$ is the unique Nash equilibrium of $\bar{\Gamma}^{x_2}$, the unique subgame perfect equilibrium of the whole game is $\bar{b} = (\{x_1\}, \{z_1\}, \{x_2\})$. However, the games are isomorphic, and the isomorphism r from Γ onto $\bar{\Gamma}$ given by $r(\{x_4\}) = \{z_3\}$, $r(\{x_1\}) = \{x_2\}$, $r(\{x_2\}) = \{z_1\}$, $r(\{x_3\}) = \{z_2\}$, $r(\{z_1, z_2, z_3\}) = \{x_2\}$ does not map b onto $\bar{b}$.

In contrast to subgame perfect equilibria, the stronger concept of perfect equilibria is invariant with respect to weak isomorphisms even in pathological cases.

Theorem 3.3.6.
Perfect equilibria of extensive games as a solution concept are invariant with respect to weak isomorphisms.

Proof. Let Γ and $\bar{\Gamma}$ be extensive games and let r be an isomorphism from Γ onto $\bar{\Gamma}$. Let further be b a perfect equilibrium of Γ. By Selten (1975, Theorem 4), b is also a perfect equilibrium of Γ's agent normal form. Let $\mathbf{r}$ be induced by r via (3.14). Since $\mathbf{r}$ is also the bijection induced by some isomorphism from Γ's agent normal form onto $\bar{\Gamma}$'s agent normal form (see the proof of Theorem 3.3.2), by Corollary 2.3.2, $\mathbf{r}(b)$ is a perfect equilibrium of $\bar{\Gamma}$'s agent normal form. By Selten (1975, Theorem 4), again, $\mathbf{r}(b)$ is a perfect equilibrium of $\bar{\Gamma}$. Since the converses of isomorphisms are isomorphisms, this proves the claim. $\square$

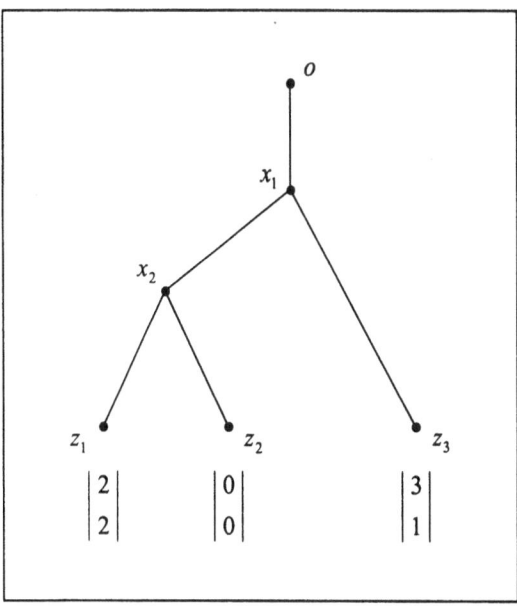

Fig. 3.11. Subgame perfection is not invariant with respect to isomorphisms #2

Since perfect equilibria of extensive games can be defined via their agent normal form, this result is not too astonishing. Kreps & Wilson's (1982) weaker concept of sequential equilibria (see Definition 3.4.1, p. 83) directly refers to the sequence of moves. However, weak isomorphisms always carry sequential equilibria into sequential equilibria.

Theorem 3.3.7.
Sequential equilibria as a solution concept are invariant with respect to weak isomorphisms.

Proof. Let Γ and $\bar{\Gamma}$ be extensive games and let r be an isomorphism from Γ onto $\bar{\Gamma}$. Let further (μ, b) be a sequential equilibrium of Γ. Since (μ, b) is a sequential equilibrium, it is consistent, i.e., there is sequence $(\mu(b^n), b^n)_{n \in \mathbb{N}}$ with $b^n \in B^0$ and $\lim_{n \to \infty}(\mu(b^n), b^n) = (\mu, b)$. Consider the sequence $(\bar{\mu}(\mathbf{r}(b^n)), \mathbf{r}(b^n))_{n \in \mathbb{N}}$. Note, by (3.15), $\mathbf{r}(b^n) \in \bar{B}^0$ holds. Since $\mathbf{r}$ and $\bar{\mu}$ are continuous in b and $\bar{b}$, respectively, we have $\lim_{n \to \infty}(\bar{\mu}(\mathbf{r}(b^n)), \mathbf{r}(b^n)) = (\bar{\mu}, \mathbf{r}(b))$ for some system of beliefs $\bar{\mu}$. Hence, $(\bar{\mu}, \mathbf{r}(b))$ is consistent. It remains to show, that $(\bar{\mu}, \mathbf{r}(b))$ is sequentially rational.

First, some auxiliary propositions are proven. We have

$$\mathbf{r}^+(\mathbf{A}^+(h)) = \bar{\mathbf{A}}^+(\nu(h)) \qquad (3.17)$$

for all $h \in H$. Since $\mathbf{A}^+(h) = \{\mathbf{a}^+ | a(\psi(z(\mathbf{a}^+))) \cap A_h \neq \emptyset\}$, this can be seen from the following chain of equations which can be read in both directions,

$$r(a(\psi(z(\mathbf{a}^+)))\cap A_h) = r(a(\psi(z(\mathbf{a}^+))))\cap r(A_h)$$
$$= \bar{a}(\bar{\psi}(\theta(z(\mathbf{a}^+))))\cap \bar{A}_{\nu(h)}$$
$$= \bar{a}(\bar{\psi}(\bar{z}(\mathbf{r}^+(\mathbf{a}^+))))\cap \bar{A}_{\nu(h)}$$

where the single equations follow from r's bijectivity, **PTH** and **ISA**, and Lemma 3.3.1 (p. 69), respectively. By (3.4), (3.9), and (3.10), we have

$$u_x(b) := \sum_{\mathbf{a}^+\in A^+(x)} \frac{\text{prob}(\mathbf{a}^+|b)}{\sum_{\mathbf{a}^{+\prime}\in A^+(x)} \text{prob}(\mathbf{a}^{+\prime}|b)} u_{i(x)}(\mathbf{a}^+)$$

for all $b\in B^0$ where $\mathbf{A}^+(x)$ denotes the set of all extended action combinations that lead to terminal nodes that precede x. Hence, by (3.8) and (3.11), we have

$$u_h(\mu(b),b) = \sum_{\mathbf{a}^+\in A^+(h)} \frac{\text{prob}(\mathbf{a}^+|b)}{\sum_{\mathbf{a}^{+\prime}\in A^+(h)} \text{prob}(\mathbf{a}^{+\prime}|b)} u_{i(h)}(\mathbf{a}^+) \qquad (3.18)$$

for all $b\in B^0$ and h. Furthermore, the beliefs $\mu(b)$ derived from b by Bayes' rule at some information set h are not affected by changing player $i(h)$'s part of b only: By perfect recall, any action of $i(h)$ that precedes some $x\in h$ also precedes all other nodes of h. Therefore, changing $i(h)$'s behavior strategy only changes the $\text{prob}(x|b)$ of all $x\in h$ by the same factor, and $\mu(x)$ remains unchanged. So we have

$$u_h(\mu(b),b) = u_h(\mu(b'),b) \qquad (3.19)$$

for all $b,b'\in B^0$ with $b'_{i'} = b_{i'}$ for all $i'\neq i(h)$. We also have

$$\bar{u}_{\nu(h)}(\bar{\mu}(\mathbf{r}(b)),\mathbf{r}(b)) = \alpha_{i(h)}u_h(\mu(b),b) + \beta_{i(h)} \qquad (3.20)$$

for all $h\in H\backslash H_{i_0}$ and $b\in B^0$. This can be seen from the following chain of equations

$$\bar{u}_{\nu(h)}(\bar{\mu}(\mathbf{r}(b)),\mathbf{r}(b)) = \sum_{\bar{\mathbf{a}}^+\in\bar{A}^+(\nu(h))} \frac{\text{prob}(\bar{\mathbf{a}}^+|\mathbf{r}(b))}{\sum_{\bar{\mathbf{a}}^{+\prime}\in\bar{A}^+(\nu(h))} \text{prob}(\bar{\mathbf{a}}^{+\prime}|\mathbf{r}(b))} \bar{u}_{\bar{i}(\nu(h))}(\bar{\mathbf{a}}^+)$$

$$= \sum_{\mathbf{a}^+\in A^+(h)} \frac{\text{prob}(\mathbf{r}^+(\mathbf{a}^+)|\mathbf{r}(b))}{\sum_{\mathbf{a}^{+\prime}\in A^+(h)} \text{prob}(\mathbf{r}^+(\mathbf{a}^{+\prime})|\mathbf{r}(b))} \bar{u}_{\pi(i(h))}(\mathbf{r}^+(\mathbf{a}^+))$$

$$= \sum_{\mathbf{a}^+\in A^+(h)} \frac{\text{prob}(\mathbf{a}^+|b)}{\sum_{\mathbf{a}^{+\prime}\in A^+(h)} \text{prob}(\mathbf{a}^{+\prime}|b)} \bar{u}_{\pi(i(h))}(\mathbf{r}^+(\mathbf{a}^+))$$

$$= \sum_{\mathbf{a}^+\in A^+(h)} \frac{\text{prob}(\mathbf{a}^+|b))}{\sum_{\mathbf{a}^{+\prime}\in A^+(h)} \text{prob}(\mathbf{a}^{+\prime}|b)} (\alpha_{i(h)}u_{i(h)}(\mathbf{a}^+) + \beta_{i(h)})$$

$$= \alpha_{i(h)}u_h(\mu(b),b) + \beta_{i(h)}$$

which follow from (3.18), v's bijectivity and (3.17), Lemma 3.3.2, Lemma 3.3.3, and (3.18), respectively.

Suppose now, on the contrary, $(\bar{\mu}, \mathbf{r}(b))$ were not sequential rational. Then, some $\bar{i} \in \bar{I} \backslash i_0$, $\bar{h} \in \bar{H}_{\bar{i}}$ and $\bar{b} \in \bar{B}$ exist such that $\bar{b}_{\bar{h}'} = \mathbf{r}_{\bar{h}'}(b)$ for all $\bar{h}' \notin \bar{H}_{\bar{i}}$ and $\bar{u}_{\bar{h}}(\bar{\mu}, \bar{b}) > \bar{u}_{\bar{h}}(\bar{\mu}, \mathbf{r}(b))$. Let $(\bar{b}^n)_{n \in \mathbb{N}}$ be some sequence such that $\lim_{n \to \infty} \bar{b}^n = \bar{b}$, $\bar{b}^n \in \bar{B}^0$, and $\bar{b}^n_{\bar{h}'} = \mathbf{r}_{\bar{h}'}(\bar{b}^n)$ for all $\bar{h}' \notin \bar{H}_{\bar{i}}$. Since $\bar{u}_{\bar{h}}$ is continuous in both arguments, there is some $N \in \mathbb{N}$ and some $\varepsilon \in \mathbb{R}$, $\varepsilon > 0$ such that

$$\bar{u}_{\bar{h}}(\bar{\mu}(\mathbf{r}(b^n)), \bar{b}^n) > \bar{u}_{\bar{h}}(\bar{\mu}(\mathbf{r}(b^n)), \mathbf{r}(b^n)) + \varepsilon$$

for all $n > N$. Since $\bar{b}^n$ only differs from $\mathbf{r}(b^n)$ at information sets of player $\bar{i}(\bar{h})$, by (3.19), we have

$$\bar{u}_{\bar{h}}(\bar{\mu}(\bar{b}^n), \bar{b}^n) > \bar{u}_{\bar{h}}(\bar{\mu}(\mathbf{r}(b^n)), \mathbf{r}(b^n)) + \varepsilon$$

and, by (3.20),

$$u_{\nu^{-1}(\bar{h})}(\mu(\mathbf{r}^{-1}(\bar{b}^n)), \mathbf{r}^{-1}(\bar{b}^n)) > u_{\nu^{-1}(\bar{h})}(\mu(b^n), b^n) + \frac{\varepsilon}{\alpha_{i(\nu^{-1}(\bar{h}))}}$$

Since by **PL** b^n and $\mathbf{r}^{-1}(\bar{b}^n)$ differ at information sets of player $i(\nu^{-1}(\bar{h}))$ only, by (3.19) we have

$$u_{\nu^{-1}(\bar{h})}(\mu(b^n), \mathbf{r}^{-1}(\bar{b}^n)) > u_{\nu^{-1}(\bar{h})}(\mu(b^n), b^n) + \frac{\varepsilon}{\alpha_{i(\nu^{-1}(\bar{h}))}}$$

Taking the limits, we have

$$u_{\nu^{-1}(\bar{h})}(\mu, \mathbf{r}^{-1}(\bar{b})) > u_{\nu^{-1}(\bar{h})}(\mu, b)$$

where $\mathbf{r}^{-1}(\bar{b})$ differs from b at information sets of player $i(\nu^{-1}(\bar{h}))$ only. Clearly, this contradicts (μ, b) being sequentially rational. Thus, $(\bar{\mu}, \mathbf{r}(b))$ is a sequential equilibrium. Since the converses of isomorphisms are isomorphism again, this proves the claim. $\square$

In view of Definition 3.3.5, weak isomorphisms of extensive games respect the order of moves to the essential extent. The Theorems 3.3.4 and 3.3.5 complement these findings.

3.4 Strong Isomorphisms of Extensive Games

The following notion of strong isomorphisms of extensive games is a slight extension of Peleg et al.'s (1999) isomorphisms: They do not allow for positive affine transformations of the utility functions, i.e., they fix all α_i at 1 and all β_i at 0.

Definition 3.4.1 (Strong isomorphisms of extensive games).
Let $\Gamma = (T, \lhd, I, P, H, A, p, u)$ and $\bar{\Gamma} = (\bar{T}, \bar{\lhd}, \bar{I}, \bar{P}, \bar{H}, \bar{A}, \bar{p}, \bar{u})$ be extensive games. A strong isomorphism from Γ onto $\bar{\Gamma}$ is a bijection $f : T \to \bar{T}$ with the following properties: There are bijections $r : A \to \bar{A}$, $\nu : H \to \bar{H}$, $\pi : I \to \bar{I}$ such that

TR	$x \lhd x'$ implies $f(x) \bar{\lhd} f(x')$ for all $x, x' \in T$,
AC	$r(a) = f(a)$ for all $a \in A$,
IS	$f(h) = \nu(h)$ for all $h \in H$,
ISA	$r(A_h) = \bar{A}_{\nu(h)}$ for all $h \in H$,
P	$f(P_i) = \bar{P}_{\pi(i)}$ for all $i \in I$,
CP	$f(P_{i_0}) = \bar{P}_{i_0}$,
CPR	$p_h(a) = \bar{p}_{\nu(h)}(r(a))$ for all $h \in H_{i_0}$ and $a \in A_h$,
PY*	there are constants $\alpha_i, \beta_i \in \mathbb{R}$, $\alpha_i > 0$ for all $i \in I \backslash i_0$ such that

$$\bar{u}_{\pi(i)}(f(z)) = \alpha_i u_i(z) + \beta_i$$

for all $z \in Z$ and $i \in I \backslash i_0$.

Condition **TR** (short for **TR**ee structure)—respecting $\lhd$—simply says that trees of isomorphic games have to be isomorphic. And it is this requirement which—in interaction with **AC**—makes these isomorphisms 'strong'.

Condition **AC** (short for **AC**tion structure) is equivalent to requiring that nodes from Γ should be mapped onto the same action in $\bar{\Gamma}$ if and only if they belong to the same action in Γ. Thereby, a unique mapping r is determined. This way, the mapping r is connected with the mapping f in an intuitive and straightforward way. Note that **AC**—together with **ISA** and **IS** (short for Information Set structure)—constitutes Peleg et al.'s (1999) requirement of respecting the choice partition, where **IS** itself is required as respecting the information partition.

However, the definition above is a bit redundant: Since $V(a) = h$ for all $a \in A_h$, **TR** and **AC** imply **IS** and **ISA**. Additionally, in presence of **TR** and **AC**, **P** (short for Player structure) and **CP** (short for Chance moves' Probabilities) are equivalent to **PL** and **CPL**, respectively. In view of **PY***, **PY** also is met by strong isomorphisms—since **TR** implies $f(Z) = \bar{Z}$, the restriction of f to Z constitutes required bijection $\theta : Z \to \bar{Z}$. Finally, **TR** implies $f(\psi(x)) = \bar{\psi}(f(x))$ for all $x \in X$, and therefore, together with **AC**, also **PTH** and **PTH**[+]. So we have the following corollary.

Corollary 3.4.1.
Let f be a strong isomorphism from Γ onto $\bar{\Gamma}$. Then, the bijection $r : A \to \bar{A}$ determined by r according to **AC** is a weak isomorphism from Γ onto $\bar{\Gamma}$ that satisfies **PTH**[+].

In order to establish the relation between weak and strong isomorphisms, some properties of strengthened condition **PTH**—**PTH**[+]—are explored. As the theorem below shows, **PTH**[+] has very strong implications indeed.

Theorem 3.4.1.
Let Γ and $\bar{\Gamma}$ be extensive games, and let $r : A \to \bar{A}$ be a bijection that satisfies $\boldsymbol{PTH^+}$. Then, a unique bijection $f : T \to \bar{T}$ exists such that

TR $x \lhd x'$ *implies* $f(x) \bar{\lhd} f(x')$ *for all* $x, x' \in T$,
AC $r(a) = f(a)$ *for all* $a \in A$.

Proof. Let θ be the unique bijection $\theta : Z \to \bar{Z}$ induced by r according to $\boldsymbol{PTH^+}$. Consider the mapping $f : T \to \bar{T}$,

$$f(x) := \begin{cases} \bar{o} & : x = o, \\ x' & : x \neq o, \exists z \in Z : x \in \psi(z), x' \in r(a(x)) \cap \bar{\psi}(\theta(z)). \end{cases} \tag{3.21}$$

Since $(T, \lhd)$ is a tree, there is a $z \in Z$ such that $x \in \psi(z)$. By $\boldsymbol{PTH^+}$ the set $r(a(x)) \cap \bar{\psi}(\theta(z))$ is not empty for $x \neq o$, and since the game has perfect recall, it has a cardinality of one. Further, it is shown that for another $z' \in Z$ with $x \in \psi(z')$, we have $r(a(x)) \cap \bar{\psi}(\theta(z)) = r(a(x)) \cap \bar{\psi}(\theta(z'))$. Suppose on the contrary that $\bar{x} \neq \bar{x}'$ existed where $\{\bar{x}\} := r(a(x)) \cap \psi(\theta(z))$ and $\{\bar{x}'\} := r(a(x)) \cap \bar{\psi}(\theta(z'))$. Then, there were nodes $\bar{y}, \bar{y}' \in \bar{T}$ such that $\bar{y} \in \bar{\psi}(\bar{x})$, $\bar{y} \notin \bar{\psi}(\bar{x}')$, $\bar{y}' \notin \bar{\psi}(\bar{x})$, $\bar{y}' \in \bar{\psi}(\bar{x}')$, and $\bar{V}(\bar{y}) = \bar{V}(\bar{y}')$. Clearly, $\bar{a}(\bar{y}) \neq \bar{a}(\bar{y}')$. Since the game has perfect recall, we have $\bar{a}(\bar{y}) \cap \bar{\psi}(\bar{x}') = \emptyset$ and $\bar{a}(\bar{y}') \cap \bar{\psi}(\bar{x}) = \emptyset$. But $\boldsymbol{PTH^+}$ implies that $\{\bar{a}(w) | w \in \bar{\psi}(\bar{x})\} = \{\bar{a}(w) | w \in \bar{\psi}(\bar{x}')\}$, a contradiction. Thus, f is well-defined.

Let $x, x' \in T$ and $z, z' \in Z$ such that $x \neq x'$, $x \in \psi(z)$ and $x' \in \psi(z')$. Then, either $a(x) \neq a(x')$ or $z \neq z'$. If $a(x) \neq a(x')$ then $f(x) \neq f(x')$ by definition. Suppose $a(x) = a(x')$ and $f(x) = f(x')$. Since $(T, \lhd)$ is a tree, there are $y, y' \in T$ such that $y \in \psi(x)$, $y \notin \psi(x')$, $y' \in \psi(x')$, $y' \notin \psi(x)$, and $V(y) = V(y')$. Clearly, $a(x) \neq a(x')$. Since the game has perfect recall, we have $a(y) \cap \psi(x') = \emptyset$ and $a(y') \cap \psi(x) = \emptyset$. Since r is bijective, $\boldsymbol{PTH^+}$ then implies $r(\{\bar{a}(w) | w \in \bar{\psi}(f(x))\}) \neq r(\{\bar{a}(w) | w \in \bar{\psi}(f(x'))\})$, a contradiction. Hence, f is injective. Since both r and θ are bijective, f is surjective. Thus, f is bijective.

By (3.21), condition **AC** follows immediately. Consider any $x, x' \in T$. Suppose $x \lhd x'$. Then there is a z such that $x, x' \in \psi(z)$. By (3.21), we have $f(x), f(x') \in \bar{\psi}(\theta(z))$. By $\boldsymbol{PTH^+}$ and **AC**, $f(x) \bar{\lhd} f(x')$ follows. Thus, **TR** holds.

Consider another bijection $f' : T \to \bar{T}$ that meets **AC** and **TR**. By **AC**, we have $f'(o) = \bar{o}$. For any $x \in T \backslash o$ there is a $z \in Z$ such that $x \in \psi(z)$. By **AC**, we have $f(x), f'(x) \in r(a(x))$, and $\boldsymbol{PTH^+}$ implies $f(x), f'(x) \in \bar{\psi}(\theta(z))$. In view of Remark 3.2.1, $f(x)$ and $f'(x)$ coincide. Thus, $f = f'$—f is unique. □

Corollary 3.4.2.
Let r be a weak isomorphism from Γ onto $\bar{\Gamma}$ that satisfies $\boldsymbol{PTH^+}$. Then, the bijection $f : T \to \bar{T}$ constructed according to (3.21) is a strong isomorphism from Γ onto $\bar{\Gamma}$.

Proof. According to Theorem 3.4.1, f satisfies **TR** and **AC** which in turn, as shown above, imply **IS** and **ISA**. By **PL** and **CPL**, **P** and **CP** are met as well. Since **AC** and **TR** imply $r(a(\psi(z))) = \bar{a}(\bar{\psi}(f(z)))$, by **PTH$^+$**, we have $f(z) = \theta(z)$. Hence, **PY** and **PY*** coincide. $\square$

Corollaries 3.4.1 and 3.4.2 together show that strong isomorphisms and weak isomorphisms obeying **PTH$^+$** are equivalent. Thus, the difference between both concepts does not lie in the fact that one of them is based on bijections of the action partitions and the other is based on bijections of the node set, but the key difference lies in the underlying path conditions: While weak isomorphisms have to respect (unordered) paths expressed by actions only—condition **PTH**, strong isomorphisms are bound to preserve ordered paths.

3.5 Traditional Extensive Representations of Strategic Games

While in strategic games the players are considered to "move" simultaneously, in extensive games, there is some order of the players' moves. Thus, the traditional extensive representations of strategic games discriminate between the players—one of them moves first, another one as the second and so on. For example, Figure 3.7 shows a traditional extensive representation of the strategic game in Figure 2.5.

In the following, the basic properties of ordering extensive representations are sketched: Consider a strategic game G and some order of the players $\jmath$ given by a bijection $\jmath : \{1, 2, \dots |I|\} \to I$. The intended interpretation of $\jmath$ is that $\jmath(1)$ is the first player (to move), $\jmath(2)$ the second one, and so on. The extensive game representation of G with respect to the players' order $\jmath$— or the $\jmath$-ordered representation is denoted $\Gamma^\jmath(G) = (T^\jmath, \lhd^\jmath, I^\jmath, P^\jmath, H^\jmath, A^\jmath, p^\jmath, u^\jmath)$. It has the following properties: $Z = \{z^\jmath[s] | s \in S\}$, $I = I^\jmath$, $P^\jmath = H^\jmath$, $A^\jmath_i = \{a^\jmath[s_i] | s_i \in S_i\}$ for all i, $p^\jmath = \emptyset$, $u^\jmath_i(z^\jmath[s]) = u_i(s)$ for all $s \in S$, $a^\jmath(\psi^\jmath(z^\jmath[s])) = \{a^\jmath[s_i] | i \in I\}$. A full definition is given in the next subsection.

3.5.1 Definition

Let $G = (I, (S_i)_{i \in I}, (u_i)_{i \in I})$ be a finite strategic game and let $\jmath$ be an order of I. The extensive game $\Gamma^\jmath(G) = (T^\jmath, \lhd^\jmath, I^\jmath, P^\jmath, H^\jmath, A^\jmath, p^\jmath, u^\jmath)$ is called the extensive form representation of G with respect to the players' order $\jmath$—or the $\jmath$-ordered representation. Its constituents are as follows:

The tree $(T^\jmath, \lhd^\jmath)$ is defined inductively: We set $T^\jmath_0 := \{\{\jmath\}\}$, $T^\jmath_n := \{x | \exists x' \in T_{n-1}, s_{\jmath(n)} \in S_{\jmath(n)} : x = x' \cup \{s_{\jmath(n)}\}\}$ for all $n \in \{1, 2, \dots |I|\}$ and $T^\jmath := \bigcup_{n \in \{0,1,2,\dots|I|\}} T^\jmath_n$. For all $x, x' \in T^\jmath$, $x \lhd^\jmath x'$ holds if and only if $x \neq x'$ and $x \subseteq x'$. This implies that the root is $o^\jmath = \{\jmath\}$ and that the set of terminal nodes is $Z^\jmath = T^\jmath_{|I|}$. Any node $x \in T^\jmath_n$ can be interpreted as being reached after the move of the nth player according to the order $\jmath$. It comprises $\jmath$ itself and exactly those strategies that have been chosen by these n players. Thus, $Z^\jmath$ contains exactly one terminal node

$$z^\jmath[s] := \{s_i | i \in I\} \cup \{\jmath\} \tag{3.22}$$

for any strategy combination s in G—the node containing just the strategies that actually have been chosen by a strategy combination.

We set $I^\jmath = I$. The player partition $P^\jmath$ is defined as $\{P_i^\jmath | i \in I\}$ where $P_i^\jmath := T_{\jmath^{-1}(i)-1}$ for all $i \in I$. Since T_n contains the nodes that are reached after the nth player's move, T_{n-1} contains player n's decision nodes. Clearly, $P_{i_0}^\jmath = \emptyset$ and $p^\jmath = \emptyset$—in $\Gamma^\jmath$, there are no chance moves. Since the players are supposed to move once only, $H^\jmath := \bigcup_{i \in I^\jmath} H_i^\jmath$ where $H_i^\jmath := \{P_i^\jmath\}$ for all $i \in I^\jmath$. Thus, $H^\jmath = P^\jmath$.

Each strategy s_i of G is represented by all nodes that are reached by it immediately; we set $a^\jmath[s_i] := \{x | x \in T_{\jmath^{-1}(i)}^\jmath, s_i \in x\}$ for all $i \in I$ and $s_i \in S_i$. Further we set $A_i^\jmath = A_{\{P_i^\jmath\}}^\jmath := \{a^\jmath[s_i] | s_i \in S_i\}$ for all $i \in I^\jmath$ and $A^\jmath := \bigcup_{i \in I^\jmath} A_i^\jmath$. Therefore, we have

$$a^\jmath(\psi^\jmath(z^\jmath[s])) = \{a^\jmath[s_i] | i \in I\} \tag{3.23}$$

for all $s \in S$. Finally, we define

$$u_i^\jmath(z^\jmath[s]) := u_i(s) \tag{3.24}$$

for all $s \in S$ and $i \in I$.

Note that this definition is very similar to the sequence representation of extensive games (see Osborne & Rubinstein 1994). Besides the fact that the terminal node is not empty, the nodes are represented by sets instead of sequences. Since the order of players is fixed and the same for all game paths, this difference is inessential.

3.5.2 Symmetry in Traditional Representations of Strategic Games

In making up an ordering extensive game representation of strategic games as described above, there is some degree of arbitrariness in choosing the order $\jmath$. As the following theorems show, this does not affect the relation between isomorphisms and symmetries of strategic games on the one hand, and weak isomorphisms and symmetries of extensive games on the other. In view of the above characterization of ordering representations, the proof is quite immediate—consider the mapping $r : A^\jmath \to A^{\bar{\jmath}}$, $r(a^\jmath[s_i]) = a^{\bar{\jmath}}[s_i]$ for all $i \in I$ and $s_i \in S_i$. Note that it is crucial for this result, that, according to condition **PTH**, game paths have to be mapped onto each other only, i.e., the order of actions within a game path is irrelevant. Since the players' order varies between different ordering representations, these representations need not to be strongly isomorphic.

Theorem 3.5.1.
Let G be a finite strategic game and let $\jmath$ and $\bar{\jmath}$ be orders of I. Then, its ordered extensive game representations $\Gamma^{\jmath}(G)$ and $\Gamma^{\bar{\jmath}}(G)$ are weakly isomorphic.

The next theorem reveals that weak isomorphisms of ordering extensive representations are compatible with isomorphisms of strategic games: Any symmetry $(\pi, (r_i)_{i \in I})$ of G induces a symmetry r of $\Gamma^{\jmath}(G)$ with $r(a^{\jmath}[s_i]) = a^{\jmath}[r_i(s_i)]$ for all $i \in I$ and $s_i \in S_i$. Conversely, any symmetry r of $\Gamma^{\jmath}(G)$ induces a symmetry $(\pi, (r_i)_{i \in I})$ of G with $\pi(i) = \pi^{\jmath}(i)$ for all $i \in I$ and $r_i(s_i) = s_{i'}$ iff $r(a^{\jmath}[s_i]) = a^{\jmath}[s_{i'}]$ for all $i, i' \in I$, $s_i \in S_i$, and $S_{i'}$.

Theorem 3.5.2.
Strategies in a finite strategic game are symmetric if and only if their representations are weakly symmetric in some ordered extensive game representation of this game.

Proof. Let $G = (I, (S_i)_{i \in I}, (u_i)_{i \in I})$ be a finite strategic game and let $\Gamma^{\jmath}(G)$ be the $\jmath$-ordered extensive game representation of G for some order $\jmath$ of I.

Let $\mathbf{f} = (\pi, (r_i)_{i \in I})$ be a symmetry of G. Consider the mapping $r : A^{\jmath} \to A^{\jmath}$ with $r(a^{\jmath}[s_i]) = a^{\jmath}[r_i(s_i)]$ for all $i \in I$ and $s_i \in S_i$. Obviously, r is bijective and induces the following bijections according to Definition 3.3.1 (p. 67): **(ISA)** $\nu^{\jmath} : H^{\jmath} \to H^{\jmath}$ such that $\nu^{\jmath}(P_i^{\jmath}) := P_{\pi(i)}^{\jmath}$ for all $P_i \in H^{\jmath}$, **(PL)** $\pi^{\jmath} : I^{\jmath} \to I^{\jmath}$ such that $\pi^{\jmath}(i) := \pi(i)$ for all $i \in I$; and, **(PTH)** $\theta^{\jmath} : Z^{\jmath} \to Z^{\jmath}$ such that $\theta^{\jmath}(z^{\jmath}[s]) := z^{\jmath}[f(s)]$ for all $z^{\jmath}[s] \in Z^{\jmath}$ where f is given by $\mathbf{f}$ via (2.6). The latter can be seen from

$$
\begin{aligned}
a^{\jmath}(\psi(\theta^{\jmath}(z^{\jmath}[s]))) &= a^{\jmath}(\psi^{\jmath}(z^{\jmath}[f(s)])) && \text{def. } \theta^{\jmath} \text{ above,} \\
&= \{a^{\jmath}(f_i(s)) | i \in I\} && (3.23), \\
&= \{a^{\jmath}(r_i(s_i)) | i \in I\} && (2.6), \\
&= \{r(a^{\jmath}[s_i]) | i \in I\} && \text{def. } r \text{ above,} \\
&= r(a^{\jmath}(\psi(z^{\jmath}[s]))) && (3.23).
\end{aligned}
$$

In view of

$$
\begin{aligned}
u_{\pi^{\jmath}(i)}^{\jmath}(\theta^{\jmath}(z^{\jmath}[s])) &= u_{\pi^{\jmath}(i)}^{\jmath}(z^{\jmath}[f(s)]) && \text{def. } \theta^{\jmath} \text{ above,} \\
&= u_{\pi(i)}^{\jmath}(z^{\jmath}[f(s)]) && \text{def. } \pi \text{ above,} \\
&= u_{\pi(i)}(f(s)) && (3.24), \\
&= \alpha_i u_i(s) + \beta_i && (2.5), \\
&= \alpha_i u_{i\jmath}^{\jmath}(z^{\jmath}[s]) + \beta_i && (3.24),
\end{aligned}
$$

condition **PY** holds. Thus, r is a weak symmetry of $\Gamma^{\jmath}(G)$ that, in addition, transfers the symmetry of strategies in G onto their representations in $\Gamma^{\jmath}(G)$.

Let r be a symmetry of $\Gamma^{\jmath}(G)$, and let $\pi^{\jmath}$, $\nu^{\jmath}$, and $\theta^{\jmath}$ the bijections induced by r according to Definition 3.3.1. By **PL**, a bijection $\pi : I \to I$ is induced by r with $\pi(i)^j = \pi^j(i^j)$ for all $i \in I$. By **ISA**, for all $i \in I$, bijections $r_i : S_i \to S_{\pi(i)}$ are induced by r with $a^j(r_i(s_i)) = r(a^j[s_i])$ for all $s_i \in S_i$. We have

$$
\begin{aligned}
a^j(\psi(\theta^j(z^j[s]))) &= r(a^j(\psi(z^j[s]))) && \textbf{PTH,} \\
&= r(\{a^j[s_i]|i \in I\}) && (3.23) \\
&= \{r(a^j[s_i])|i \in I\} && \\
&= \{a^j(r_i(s_i))|i \in I\} && \text{def. } r_i \text{ above,} \\
&= \{a^j(f_i(s))|i \in I\} && (2.6), \\
&= a^j(\psi^j(z^j[f(s)])) && (3.23) \\
\theta^j(z^j[s]) &= z^j[f(s)] && \text{Rem. 3.2.1,}
\end{aligned}
$$

for all $s \in S$. Therefore,

$$
\begin{aligned}
u_{\pi(i)}(f(s)) &= u^j_{\pi(i)^j}(z^j[f(s)]) && (3.24), \\
&= u^j_{\pi^j(i^j)}(z^j[f(s)]) && \text{def. } \pi \text{ above,} \\
&= u^j_{\pi^j(i^j)}(\theta^j(z^j[s])) && \text{see above,} \\
&= \alpha_{ij} u^j_{ij}(z^j[s]) + \beta_{ij} && \textbf{PY,} \\
&= \alpha_{ij} u_i(s) + \beta_{ij} && (3.24),
\end{aligned}
$$

i.e., condition (2.5) (p. 17) holds for all $i \in I$ and $s \in S$. Thus, $(\pi, (r_i)_{i \in I})$ is a symmetry of G that, in addition, transfers the symmetry of actions in $\Gamma^j(G)$ onto the strategies they represent in G. $\square$

Since strong isomorphism are also weak ones, the strong symmetry of strategy representations in the ordered extensive representation is transferred to the symmetry of these strategies in the strategic game, i.e., the transformation into some ordering extensive game representation does not create new strong symmetries. On the other hand, the symmetry of different players' strategies cannot transfer to the strong symmetry of their representations in the ordered extensive representation, i.e., the transformation into some traditional extensive game representation may destroy the symmetry of strategies with respect to strong symmetry.

3.6 Conclusion

In this chapter the concept of weak isomorphisms of extensive games was introduced and advocated. Its properties as explored above can be summarized as follows:

- In non-pathological cases, the path condition **PTH** as part of weak isomorphisms is equivalent to the weakest condition—**PTH⁻**—that preserves the relation between actions and outcomes (terminal nodes)—Lemma 3.3.1 and Theorem 3.3.1.

- The main difference between weak isomorphisms and Peleg et al.'s (1999) strong isomorphisms lies in the path condition—Corollaries 3.4.1 and 3.4.2. While Peleg et al. (1999) imply a strict preservation of (ordered) game paths—condition $\mathbf{PTH^+}$, weak isomorphisms only require the preservation of unordered paths expressed by actions—condition $\mathbf{PTH}$.
- As Peleg et al. (1999) observe, their strong isomorphisms are incompatible with the traditional extensive representation of strategic games concerning the respective symmetries. In contrast, weak isomorphisms fit these representations—Theorem 3.5.2. Thus, one could argue that respecting the order of moves in the strict sense embodied in condition $\mathbf{PTH^+}$ is too strong.
- Although weak isomorphisms do not strictly preserve the order of moves, they preserve the order in relevant respects: In non-pathological cases, they respect the subgame structure—Theorem 3.3.4—and they carry subgame perfect equilibria again into subgame perfect equilibria. The latter also holds for Nash equilibria, perfect equilibria, and sequential equilibria in general. Since sequential equilibria explicitly refer to the order of moves, the transfer property seems to be especially important for this concept—Theorem 3.3.7.
- In addition, the symmetry invariance requirement based on weak isomorphisms is compatible with the equilibrium concepts under consideration: even symmetry invariant perfect equilibria do always exist—Theorem 3.3.3.

Weak isomorphisms meet the weakest requirements that one should demand from a concept of isomorphisms for (non-pathological) finite extensive games with perfect recall. As argued before, it is implicit in this notion that the sequential nature of moves is a technical peculiarity of extensive games rather than a representation of real world facts. Finally, we provide some more justification for the use of these isomorphisms by Selten (1983, 1988), Oh (1995), and Casajus (1998) as well as for its use in Chapter 4.

4 Dynamic Focal Points in Framed Extensive Games

In order to formalize dynamic focal points, the FSG approach as introduced in Chapter 2 is transferred to extensive games in this chapter. This is done (a) by extending extensive games by frames analogous to the frames of strategic games, and (b) then extending the weak isomorphisms introduced in Chapter 3 to the resulting *framed extensive games* (FEGs) by a language variance requirement analogous to the respective requirement for FSGs.

Besides the introduction of FEGs, mainly, the two examples presented in Section 4.2 are discussed. In Section 4.3, we analyze whether and to what extent other representations of dynamic choice problems can be utilized to explain dynamic focal points. As a result, a transfer of the FSG approach to extensive games seems to be desirable. Further, it is demonstrated that the dynamic focal points in the two examples can be supported within the FEG framework (Section 4.5).

4.1 Introduction

Schelling (1960) already suggests that focal points may rest upon precedence, i.e., the players' choice in earlier stages/occurrences of the game. Kramarz (1996) coins the term of dynamic focal points for this kind of focal points. Since (to the extent that) previous choices do not affect the preferences at later stages of the game, precedence should be viewed as providing an (additional) non-strategic description of the players' actions in these later stages (Lewis 1969). Thus, precedence is just one way of framing a game that possibly breaks the symmetry of strategies/actions. But in contrast to the static focal points considered in Chapter 2, the players' framing of the game is not given from the beginning. Frames arise or are—consciously or not—modified in the course of (and by) playing the game.

Nevertheless, at a given time, this description is given to the players. So, one could say that static focal points are important components of the dynamic ones. What the dynamic perspective adds is that the players may anticipate how present actions influence their future descriptions of the game, and thereby influence their future actions and the resulting payoffs. In turn, this foresight may lead the players to take actions that produce a favorable description. As Crawford & Haller (1990) point out, this might lead the players to take some dominated stage-game strategy.

An important example of dynamic focal points are conventions—"pattern[s] of behavior that [are] customary, expected, and self-enforcing. Everyone conforms, everyone expects the others to conform, everyone wants to conform given that everyone else conforms" (Young 1993, p. 57). More precisely, conventions can be viewed as dynamic focal points inasmuch they arise by gradual accretion of precedent (see Young 1996). Consider the problem on which side of a road to drive. In the above sense, the rule of driving on the right side is a convention. Suppose that at some encounter, both drivers keep right. Note that this does not change the players' preferences—they are solely interested in driving on the 'same' side of the road. But at subsequent encounters, driving on the right side may have become the salient or prominent option which therefore is taken on the 'rational' basis of some kind of focal point reasoning. Or, more general, the side that is taken by encounting cars more frequently becomes the salient one, hence, a driver's choice. In time, driving on the right side may spread over the whole population—a convention emerges which, as it happened frequently, is finally fixed by law.

Goyal & Janssen (1996) argue that dynamic focal points cannot be explained by individual rationality considerations alone. They adopt Agassi's (1960, p. 263) position "neither to assume the existence of all co-ordination nor to explain all of them, but rather to assume the existence of some co-ordination in order to explain the existence of some other co-ordination". In other words, we cannot explain dynamic focal points without reference to some higher-order focal points: Reconsider the example of the encounting cars. Suppose there are two cars only (e.g., on some small island) that managed to pass by driving on the right side. For the subsequent encounters, there are several optimal rules: (a) driving on the same side as at the last encounter, (b) switching the side each time they meet, (c) switching the side every nth time. Note that there are many other optimal rules. Since rule (a) is the most simple one, in a sense, it also is the salient one.

Besides the interpretation of the precedence above as some kind of framing that can be exploited 'rationally', another explanation of dynamic focal points is by bounded rationality in an evolutionary framework: The players exploit the precedent within some rule of thumb. In the example above, they simply may take the side taken by the other driver at the last encounter. This can be justified as follows: Precedent is used as an indicator for the other players' actual play to which the players respond according to their preferences. In other words, the precedent shapes the players' expectations of the other players' future play. Again, following Goyal & Janssen (1996), this cannot be justified by considerations based on individual rationality alone.

Young (1993, 1996) and Kandori, Mailath & Rob (1993), among others, show how conventions based on bounded rationality assumptions can arise in an evolutionary framework. In contrast, Crawford & Haller's (1990) and Kramarz' (1996) explanation of dynamic focal points is based on the precedent as a symmetry-breaking device. Blume & Gneezy (1998) find some support for Crawford & Haller's (1990) optimal rules for twice-repeated pure coordination games in controlled experiments.

4.2 Examples

This section presents two examples of dynamic focal points that are used throughout the chapter.

4.2.1 Crawford & Haller's Rule for Repeated Matching Games

Crawford & Haller (1990) analyze repeated two-person pure coordination games (i.e., matching games M_n, see Definition 2.3.1, p. 14) in which the players learn the other players' action of the previous stages and derive rules for optimal learning in these games, i.e., ways in which the choices at previous stages can be exploited to achieve coordination at the succeeding stages. For the case of three stage-game actions, they suggest the following optimal repeated-game strategy: At the first stage, all actions should be chosen with the same probabilities. At all following stages, a player should either repeat the coordinating stage game action of the previous stage, or in the case of discoordination, he should switch to that stage-game action that he did not use at the previous stage and that would not have lead to coordination given the other player's action. This strategy is highly efficient—it secures coordination from the second stage on. Even in case of deviation at some instance, coordination is achieved onwards.

The intuition behind this is the following: Assuming that no strategy is salient in the stage game itself, the strategies are not distinguished at the first stage of the repeated game—put formally, they are symmetric in the sense of Definition 2.3.9 (p. 19). Hence, randomization is the only symmetry invariant strategy at the first stage. But at later stages, the strategies might be distinguished by the history of the game—the fact that a stage-game strategy was chosen at the previous stages or not. I.e., by playing the game, the players, intentionally or not, produce a description of the strategies for the following stages. Now, in some sense to be made precise, the strategies need not be symmetric.

But Crawford & Haller's suggestion is just one possible way to produce and exploit a history in this game. From the third stage on, the actual play in several stages might be exploited as a coordinating device. It is not clear that players should always use the immediately previous stage only to coordinate their behavior at the following stages. Another plausible and optimal rule is to base the above decision criterion on the actual play at the first stage for any of the following stages. But Crawford & Haller's rule requires a short-term memory only, and both rules lead to the same game path—they only differ in their off-equilibrium prescriptions. A straightforward way to overcome this ambiguity is to restrict the players' memory to the previous-stage actions. This would ensure coordination even in the off-equilibrium situations. A similar result was obtained by Kramarz (1996).

For the sake of simplicity and in order to avoid the ambiguities mentioned above, a twice-repeated coordination game with three options is considered in the following.

Example 4.2.1.
There are two rounds. In each, two players independently have to choose one of three physical objects. If they take the same object, in any round, then they get a cookie; otherwise they get nothing. The objects chosen in the first round are common knowledge at the beginning of round 2.

Assuming that none of the objects is salient in some respect, the players cannot do better than randomizing in the first round. This way, the probability of coordination in the first round is $\frac{1}{3}$. In the second round, as Crawford & Haller suggest, the players can do much better—they can coordinate for sure by applying the following rule: In case of coordination in the first round, they should take the same object as in the first round, and in case of non-coordination, they should take the object that was taken by none of them in the first round.

4.2.2 Conscious Production of a Description

In the previous example, the players do not try to create a certain history—randomization guided by the symmetry of the stage-game strategies provided them a history by pure chance. Moreover, any actually played action combination at some single previous stage enables the players to coordinate for sure at all succeeding stages.

In the following example, in contrast, the players can and—as it is shown—should consciously decide on the history to provide for the later stages. Consider the twice repeated stage game without discounting and with identical payoffs for both players given as in Figure 4.1. Note that this game is quite similar to a game analyzed by Crawford & Haller (1990, p. 586).

Since the last three strategies are indistinguishable by the payoff structure, put differently, the strategies are symmetric, they should get the same probabilities at the first stage. Randomizing over these strategies leads to the stage payoff of $\frac{2}{3}$ whereas taking the payoff distinguished strategies s_{11} and s_{21} leads to immediate coordination and yields a payoff of 1. But taking the first strategy does not produce a history that distinguishes the last three strategies at the second stage. Therefore, the players' best decision at the second stage also were to take the first stage-game strategy which provides them an overall payoff of 2. As discussed in the previous example, randomizing over the last three strategies at the first stage secures coordination at

	s_{21}	s_{22}	s_{23}	s_{24}
s_{11}	1	0	0	0
s_{12}	0	2	0	0
s_{13}	0	0	2	0
s_{14}	0	0	0	2

Fig. 4.1. A stage game

the second stage. This leads to an overall payoff of $\frac{8}{3}$. Thus, it is optimal to choose the payoff-dominated symmetry invariant stage-game equilibrium in the first round in order to produce a favorable history for the second one.

4.3 Modelling Dynamic Focal Points

This section formalizes the intuition behind Example 4.2.1. Several representations are analyzed with respect to their ability to explain the focal points as incorporated in Crawford & Haller's rule. As it turns out, neither extensive games nor Oh's (1995) extension by histories allow for an explanation of dynamic focal points. In contrast, the FSG approach can be exploited to formalize dynamic focal points, even in a quite general manner. But this involves some casuistic modelling or some abuse of frames as a representation of strategic features—the sequence of moves.

4.3.1 Extensive Games

The game from Example 4.2.1 can be modelled by an extensive game $\Gamma^{CH} = (T, \lhd, I, P, H, A, p, u)$. For the sake of notational parsimony, not the full tree representation is given.

Definition 4.3.1.
The extensive game Γ^{CH} is characterized as follows:

1. $I = \{1, 2\}$,
2. $H_i = \{(i, z_1, z_2) | z_1, z_2 \in \{1, 2, 3\} \vee z_1 = z_2 = 0\}$ *for all* $i \in I$,
   $H = H_1 \cup H_2$,
3. $A_{(i, z_1, z_2)} = \{(i, z_1, z_2, z) | z \in \{1, 2, 3\}\}$ *for all* $(i, z_1, z_2) \in H$,
   $A_i = \bigcup_{(i, z_1, z_2) \in H_i} A_{(i, z_1, z_2)}$, $A = A_1 \cup A_2$,
4. $Z = \{(z_{11}, z_{12}, z_{21}, z_{22}) | z_{11}, z_{12}, z_{21}, z_{22} \in \{1, 2, 3\}\}$,
5. $a(\psi(z)) = a(\psi(z_{11}, z_{12}, z_{21}, z_{22})) = \{(1, 0, 0, z_{11}), (2, 0, 0, z_{12}), (1, z_{11}, z_{12}, z_{21}), (2, z_{11}, z_{12}, z_{22})\}$,

6. $p = \emptyset$,
7. For all $i \in I$ and $(z_{11}, z_{12}, z_{21}, z_{22}) \in Z$,

$$
u_i(z_{11}, z_{12}, z_{21}, z_{22}) = \begin{cases} 2 : z_{11} = z_{12} \wedge z_{21} = z_{22} \\[2mm] 1 : z_{11} = z_{12} \wedge z_{21} \neq z_{22} \\[2mm] 1 : z_{11} \neq z_{12} \wedge z_{21} = z_{22} \\[2mm] 0 : z_{11} \neq z_{12} \wedge z_{21} \neq z_{22} \end{cases} \tag{4.1}
$$

The intended interpretation is the following: The number $z_{ji} \in \{1, 2, 3\}$ represents player i's choice in round j; $z \in \{1, 2, 3\}$ represents some player's choice in some round. In information sets and actions, $z_i \in \{1, 2, 3\}$ represents player i's choice in round 1, where $z_i = 0$ indicates that player i has not yet made his choice, i.e., this numbering indicates the first round information sets and actions.

To be more specific, each player has an information set $(i, 0, 0)$, his first-round information set. All other information sets (i, z_{11}, z_{12}) are player i's information sets of the second round. Information set (i, z_{11}, z_{21}) is reached when the players took their actions $(1, 0, 0, z_{11})$ and $(2, 0, 0, z_{12})$ in the first round, respectively. In general, action $(i, z_{11}, z_{12}, z_{2i})$ is the z_{2i}th action of player i at his information set (i, z_{11}, z_{12}). The terminal node $z = (z_{11}, z_{12}, z_{21}, z_{22})$ is reached when the players took the actions $(1, 0, 0, z_{11})$ and $(2, 0, 0, z_{12})$ in the first round, and the actions $(1, z_{11}, z_{12}, z_{21})$ and $(2, z_{11}, z_{12}, z_{22})$ in the second round. The payoff function assigns the payoff 2 if the players got coordinated in both rounds, i.e., if they took the action with the same distinguishing number, the payoff 1 if the players got coordinated in one round only, and the payoff 0 if they got not coordinated at all. Note that it is implicit in this setting that there is no discounting.

Theorem 4.3.1.
In Γ^{CH}, there is a unique payoff-dominant weakly symmetry-invariant equilibrium b with $b_{(i,z_1,z_2)}(i, z_1, z_2, z) = \frac{1}{3}$ for all $(i, z_1, z_2, z) \in A$.

This result is not too surprising: In Γ^{CH}, all actions are different—Selten (1983, p. 287) remarks that "... the pure strategy sets of both players never coincide even if the game is obviously symmetric in any reasonable sense." Taken a step further, this means that actions of one player cannot be the same at different information sets. This easily can be seen from the fact that an action is a cell of the action partition. This way it is impossible to identify the 'same' actions in different rounds, i.e., across information sets representing the players' decisions in these rounds. Thus, the weak symmetry invariance requirement prevents the efficient outcome suggested by Crawford & Haller even in the two-rounds case.

Proof. Let the bijection $r : A \to A$ be defined by permutations q_1 and q_2 of $\{1, 2, 3\}$ and a bijection $\pi : I \to I$ such that $r(i, 0, 0, z_{1i}) = (\pi(i), 0, 0, q_1(z_{1i}))$ and $r((i, z_{11}, z_{12}, z_{2i})) = (\pi(i), q_1(z_{1\pi(1)}), q_1(z_{1\pi(2)}), q_2(z_{2i}))$ which induces the bijection $\nu : H \to H$ with $\nu(i, 0, 0) = (\pi(i), 0, 0)$ and $\nu(i, z_{11}, z_{12}) = (\pi(i), q_1(z_{1\pi(1)}), q_1(z_{1\pi(2)}))$, and the bijection $\theta : Z \to Z$ with $\theta(z) = (q_1(z_{1\pi(1)}), q_1(z_{1(\pi(2)}), q_2(z_{2\pi(1)}), q_2(z_{2\pi(2)}))$ according to Definition 3.3.1 (p. 67). It is easy to check that r is a weak symmetry of Γ^{CH}.

The symmetry r of Γ^{CH} defined by $\pi = \mathrm{id}$, $q_1(1) = 2$, $q_1(2) = 3$, $q_1(3) = 1$, and $q_2 = \mathrm{id}$ establishes the symmetry of a player's first-round actions, and the symmetry r' defined by $\pi' = \mathrm{id}$, $q_1' = \mathrm{id}$, $q_2'(1) = 2$, $q_2'(2) = 3$, and $q_2'(3) = 1$ establishes the symmetry of a player's actions at any second-round information set. Hence, in Γ^{CH}, the unique symmetry-invariant behavior-strategy combination prescribes randomization at all information sets. $\square$

4.3.2 Extensive Games with History

In order to overcome similar inefficient outcomes, Oh (1995) extends extensive games by a structure which he calls *history*. He defines "... a history $t_i(u_i)$ of player i's information set u_i by the collection of all players' past action choices, from the origin to the information set u_i which player i can identify." (p. 12). But Oh does not make the phrase "which player i can identify" more explicit. The following definition seems to meet his intentions:

Definition 4.3.2 (History of an extensive game).
Let Γ be an extensive game. A history of Γ is a correspondence hist $: H \rightrightarrows A$ *satisfying*

$$\mathrm{hist}(h) := \{a : \forall x \in h : \psi(x) \cap a \neq \emptyset\} \tag{4.2}$$

for all $h \in H$.

A history hist(h) of the information set h comprises all actions that precede h and which player $i(h)$ is able to identify in the following sense: An action a belongs to the history of some information set h, if a precedes (intersects the path of) any node of x, i.e., player $i(h)$ knows for sure that action a has been taken. Since $V(a) = h$ for all $a \in A_h$, the history of an information set h could also be considered as the history of any of its actions.

Two other interpretations seem to be conceivable:

$$\mathrm{hist}^-(h) := \{a : \exists x \in h : \psi(x) \cap a \neq \emptyset\},$$

i.e., at h, player $i(h)$ cannot exclude that some action $a \in \mathrm{hist}^-(h)$ has been taken, and

$$\mathrm{hist}^+(h) := \left\{a : \forall x' \in a : \exists x \in h : x' \in \psi(x)\right\},$$

i.e., every action $a \in \mathrm{hist}^+(h)$ necessarily leads to h. But in both cases, a player does not know for sure whether an action from the history has been taken or not.

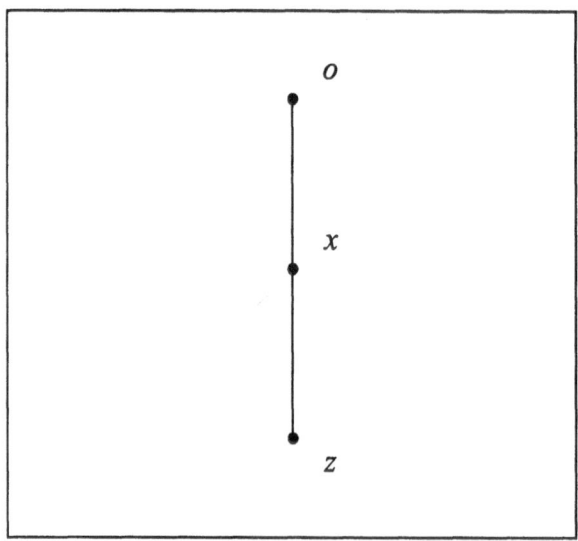

Fig. 4.2. HIS is not implied by weak isomorphisms

Example 4.3.1.
Consider the game Γ^{CH}. The histories of the first-round information sets are empty—hist$(1,0,0)$ = hist$(2,0,0)$ = $\emptyset$. As for player 1, there are no nodes, hence no actions that precede information set $(1,0,0)$. Since all first-round actions of player 1 lead to player 2's first-round information set $(2,0,0)$, none of these can be contained in its history. For quite the same reason, player 1's second-round actions never can be contained in the histories of both player's second-round information sets. So we have

$$\text{hist}(i, z_{11}, z_{12}) = \{(1,0,0,z_{11}), (2,0,0,z_{12})\}$$

for all $(i, z_{11}, z_{12}) \in H$ as any first-round combination of actions leads to a different subgame in which the players just can identify those actions that lead to the respective subgame.

In order to exploit the history of extensive games, Oh extends weak isomorphisms r by the following condition of history preservation

HIS $r(\text{hist}(h)) = \overline{\text{hist}}(\nu(h))$ for all $h \in H$.

The following example shows that **HIS** is not implied by Definition 3.3.1 (p. 67).

Example 4.3.2.
Consider the game Γ in Figure 4.2 with $T = \{o, x, z\}$, $o \lhd x \lhd z$, $I = \{1\}$, $P_1 = \{o, x\}$, $H = \{\{o\}, \{x\}\}$, and $A = \{\{x\}, \{z\}\}$. Let r be a the weak symmetry of Γ with $r(\{x\}) = (\{z\})$ and vice versa. Via **ISA**, r induces

the bijection $\nu : H \rightarrow H$ with $r(\{o\}) = (\{x\})$ and vice versa. So we have $r(\text{hist}(\{o\})) = \emptyset \neq \{\{x\}\} = \text{hist}(\{x\}) = \text{hist}(\nu(\{o\}))$, i.e., **HIS** does not hold.

The example involves a pathological game. Without this peculiarity, we have:

Theorem 4.3.2.
*In non-pathological cases, condition **HIS** is implied by weak isomorphisms.*

Firstly, a characterization of histories for non-pathological games is established.

Lemma 4.3.1.
In non-pathological cases, $a \in \text{hist}(h)$ if and only if $\mathbf{a}^+ \in \mathbf{A}^+(h)$ implies $\mathbf{a}^+_{V(a)} = a$. Note that $V(a)$ is the information set a belongs to—$a \in A_{V(a)}$.

Proof. The only-if part is immediate and even holds in pathological cases. Suppose on the contrary that $\mathbf{a}^+ \in \mathbf{A}^+(h)$ implies $\mathbf{a}^+_{V(a)} = a$ for some a, and $a \notin \text{hist}(h)$. I.e., there were some $x \in h$ such that $\psi(x) \cap a = \emptyset$. By assumption, there is some $\mathbf{a}^+$ such that $x \in \psi(z(\mathbf{a}^+))$ and $\psi(z(\mathbf{a}^+)) \cap a \neq \emptyset$. Let x' be unique element of $\psi(z(\mathbf{a}^+)) \cap a$. So we have $x \lhd x'$. Since the game is non-pathological, $N(V(x))$ contains some $x'' \neq x'$. Clearly, $a(x'') \neq a(x') = a$. But this means that there were some $\mathbf{a}^{+'} \in \mathbf{A}^+(h)$ with $\mathbf{a}^{+'}_{V(a)} = a(x'') \neq a$. A contradiction. $\square$

Proof. (Theorem 4.3.2) Let Γ and $\bar{\Gamma}$ be non-pathological extensive games, and let r be a weak isomorphism from Γ onto $\bar{\Gamma}$ which implies a bijection $\nu : H \rightarrow \bar{H}$ via **ISA**.

Let $a \in \text{hist}(h)$. By Lemma 4.3.1, $\mathbf{a}^+ \in \mathbf{A}^+(h)$ implies $\mathbf{a}^+_{V(a)} = a$. Since by (3.17) $r^+(\mathbf{A}^+(h)) = \bar{\mathbf{A}}^+(\nu(h))$, we have $\bar{\mathbf{a}}^+_{\nu(V(a))} = r(a)$ for all $\bar{\mathbf{a}}^+ \in \bar{\mathbf{A}}^+(\nu(h))$. By Lemma 4.3.1, we have $r(a) \in \overline{\text{hist}}(\nu(h))$ and therefore, $r(\text{hist}(h)) \subseteq \overline{\text{hist}}(\nu(h))$. Since inverses of weak isomorphisms also are weak isomorphisms, the converse inclusion is immediate. This establishes the claim. $\square$

Obviously, this result has fatal consequences for Oh's history approach: Condition **HIS** does not refine the concept of weak isomorphisms in any substantial extent. His main flaw is that he did not take very seriously Selten's (1983, p. 287) 'warning' quoted in Subsection 4.3.1—there are no equal actions. This is so because the game theorist's naming of the decision makers' choices as actions is considered only. Therefore, his histories together with history preservation (condition **HIS**) do not break symmetries in repeated games as intended. However, Oh's suggestion to extend Selten's (1983) symmetries of two-person extensive games into a general notion for extensive games turned out to be fruitful (see Chapter 3). Additionally, in Subsection 4.3.4, we show how Oh's histories can be used in order to get the results intended by him (see (4.4)).

s_{ij} $\mathfrak{a}$ $\ell_{i\mathfrak{a}}(s_{ij})$	s_{11}	s_{12}	s_{13}	s_{21}	s_{22}	s_{23}
me	yes	no	no	yes	no	no
other	yes	no	no	yes	no	no

Fig. 4.3. Frame for a second-round subgame after first-round coordination

s_{ij} $\mathfrak{a}$ $\bar{\ell}_{i\mathfrak{a}}(s_{ij})$	s_{11}	s_{12}	s_{13}	s_{21}	s_{22}	s_{23}
me	no	yes	no	no	no	yes
other	no	no	yes	no	yes	no

Fig. 4.4. Frame for a second-round subgame after first-round discoordination

4.3.3 Framed Strategic Games—Ad hoc Approach

Crawford & Haller's suggestion can be supported within the FSG approach.
The players' situation after first-round coordination (i.e., the resulting sub-
game) can be described by the FSG (M_3, F) with $\mathfrak{A} = \{\text{me}, \text{other}\}$, $\Lambda = \{\text{yes}, \text{no}\}$, and the label functions $\ell_{i\mathfrak{a}}$ as in Figure 4.3.

The intended interpretation of this frame is as follows: The attribute
me indicates whether the respective object was chosen by the player under
consideration—label yes—or not—label no, and the attribute other indicates which object was chosen by the respective opponent. Clearly, both
players as well as the second and the third strategy are symmetric. The lat-
ter can be seen from the symmetry $(\mathbf{f}, \mathbf{t})$ with $\mathbf{f} = (\pi, (r_i)_{i \in I})$ given by $\pi = \text{id}$
and $r_i(s_{i1}) = s_{i1}$, $r_i(s_{i2}) = s_{i3}$, and $r_i(s_{i3}) = s_{i2}$ for both i, and $\mathbf{t} = \text{id}$.
In view of Corollary 2.5.1 (p. 44), neither the second nor the third strat-
egy can be symmetric to the first one. Hence, the unique payoff-dominant
symmetry-invariant equilibrium is both players' taking the first object. After
discoordination by taking the second and the third object respectively, the
situation can be described by the FSG $(M_3, \bar{F})$ with $\mathfrak{A}$ and Λ as above, and
the label functions $\bar{\ell}_{i\mathfrak{a}}$ as in Figure 4.4.

Again, both players as well as the second and the third strategy are sym-
metric, which can be seen from the symmetry $(\mathbf{f}, \mathbf{t})$ with $\mathbf{f} = (\pi, (r_i)_{i \in I})$ given
by $\pi = \text{id}$ and $r_i(s_{i1}) = s_{i1}$, $r_i(s_{i2}) = s_{i3}$, and $r_i(s_{i3}) = s_{i2}$ for both i, and
$\mathbf{t} = (\mu, \tau)$ given by $\mu(\text{me}) = \text{other}$, $\mu(\text{other}) = \text{me}$, and $\tau = \text{id}$. Neither the
second nor the third strategy can be symmetric to the first one. Therefore,
both players taking the first object is the unique payoff-dominant symmetry-
invariant strategy combination. Also, in the case of first-round discoordination, Crawford & Haller's rule can be supported.

But in the latter case, the symmetry of strategies partly hinges on making
up the frame just in that particular way (or similarly). Consider the alterna-

$\tilde{a}$	s_{1j} $\ell_{i\tilde{a}}(s_{1j})$	s_{11}	s_{12}	s_{13}	s_{21}	s_{22}	s_{23}
first		no	yes	no	no	no	yes

Fig. 4.5. Alternative frame for a second-round subgame after first-round discoordination

tive frame $\tilde{F}$ with $\widetilde{\mathfrak{A}} = \{\texttt{first}\}$, $\varLambda = \{\texttt{yes}, \texttt{no}\}$, and the label functions $\ell_{i\tilde{a}}$ as in Figure 4.5.

The intended interpretation is the following: The attribute **first** indicates whether the respective player has taken the object in the first round or not. In contrast to frame $\bar{F}$, from Corollary 2.5.1 it follows that the second and the third strategies of the players are no longer symmetric. But the players are symmetric by the symmetry $(\mathbf{f}, \mathbf{t})$ with $\mathbf{f} = (\pi, (r_i)_{i \in I})$ given by $\pi(1) = 2$ and $\pi(2) = 1$, and $r_i(s_{11}) = s_{\pi(i)1}$, $r_i(s_{12}) = s_{\pi(i)3}$, and $r_i(s_{i3}) = s_{\pi(i)2}$ for both i, and $\mathbf{t} = \mathrm{id}$. Therefore, s_{12} and s_{23} have to get the same probabilities in a symmetry invariant strategy combination. Again, both players taking the first object is the unique payoff-dominant symmetry-invariant strategy combination.

This example shows that the symmetry of strategies is sensitive to the details of modelling the players' previous moves. Additionally, the modelling is rather ad hoc—Blume & Gneezy (1998) suggest some other framings of M_3. So it seems to be desirable to have a more general representation of extensive games as FSGs that produces the same results. The next subsection describes one way in which this can be done.

4.3.4 Framed Strategic Games—General Approach

Since it is the aim of the model to make up to identify somehow related but, of course, different actions in extensive games, at the first glance, its seems to be sufficient to consider one attribute that covers this relatedness only. Consider an extensive game $\varGamma = (T, \lhd, I, P, H, A, p, u)$ and its agent normal form $G(\varGamma) = (H \backslash H_{i_0}, (A_h)_{h \in H \backslash H_{i_0}}, (u_h)_{h \in H \backslash H_{i_0}})$ according to Definition 3.2.3 (p. 64). Let F be a one-attribute frame of $G(\varGamma)$ that formalizes the relatedness of actions as mentioned above, i.e., $|\mathfrak{A}| = 1$. Call the FSG $(G(\varGamma), F)$ the simply labelled agent normal form of $\varGamma$.

Reconsider the extensive game $\varGamma^{CH}$ from Subsection 4.3.1 and its simply labelled agent normal form $(G(\varGamma^{CH}), F^{CH})$ where the frame F^{CH} consists of $\mathfrak{A} = \{\mathtt{a}\}$, $\varLambda = \{1, 2, 3\}$, and the label functions such that $\ell_{(i, z_1, z_2)\mathtt{a}}(i, z_1, z_2, z) = z$ for all $(i, z_1, z_2, z) \in A$. I.e., across the rounds/information sets, the actions/strategies get their distinguishing number as a label with respect to the attribute $\mathtt{a}$. The intuition behind this frame is that it enables the players to identify first-round actions with the corresponding actions in the second round.

The bijections $\nu : H \to H$, $r_h : A_h \to A_{\nu(h)}$ for all $h \in H$, $\mu : \mathfrak{A} \to \mathfrak{A}$, and $\tau : \Lambda \to \Lambda$, induced by the permutation q of $\{1, 2, 3\}$ with $q(1) = 2$, $q(2) = 3$, $q(3) = 1$ according to $\nu = \mathrm{id}$,

$$r_{(i,z_1,z_2)}(i, z_1, z_2, z) = (i, z_1, z_2, q(z)) \tag{4.3}$$

for all $(i, z_1, z_2, z) \in A$, $\mu = \mathrm{id}$, $\tau(z) = q(z)$ for all $z \in \{1, 2, 3\}$ constitute a symmetry of the FSG $(G(\Gamma^{CH}), F^{CH})$: By (4.3), actions with the same distinguishing number—last component—are mapped onto each other. Since, in addition, (a) the actions are labelled by their distinguishing numbers and (b) the labelling numbers are permuted in the same way as the distinguishing numbers, condition (2.5) of Definition 2.3.6 (p. 17) and condition (2.15) of Definition 2.5.4 (p. 41) are satisfied. This means that all of one player's (information set's) actions are symmetric. Hence, we have the same result as in Subsection 4.3.1—in the unique symmetry invariant equilibrium, the players always randomize.

At closer inspection, this is not astonishing: Simply labeled agent normal forms contain no information about the sequence of related actions.

This shortcoming can be overcome by a frame

$$F_{\mathrm{hist}} = (\mathfrak{A}^*, \Lambda^*, (\ell^*_{ha^*})_{(h,a^*) \in H \backslash H_{i_0} \times \mathfrak{A}^*})$$

of the simply labelled agent normal form $(G(\Gamma), F)$. In order to formalize the sequence of related actions, the attribute set $\mathfrak{A}$ of the simply labeled agent normal form is extended by attributes a for any $a \in A$—$\mathfrak{A}^* = \mathfrak{A} \cup A$, the label set Λ is extended by the label no—$\Lambda^* = \Lambda \cup \{no\}$, and the label functions ℓ^*_{ha} for all $h \in H \backslash H_{i_0}$ and $a \in A \subset \mathfrak{A}^*$ are added. These additional label functions are defined as follows: For all $h \in H \backslash H_{i_0}$, $a \in A_h$, $h' \in H$ and $a' \in A_{h'}$ we set

$$\ell^*_{ha'}(a) = \begin{cases} \ell_{h'a}(a') : a' \in \mathrm{hist}(h) \\ no \quad\quad : a' \notin \mathrm{hist}(h) \end{cases} \tag{4.4}$$

where the $\mathrm{hist}(h)$ is defined as in (4.2). Also, we set $\ell^*_{ha} = \ell_{ha}$. The intended interpretation of the extended frame is the following: Besides its own label, for any action a, this frame 'lists' the labels of those actions of which the player knows for sure that they have been chosen before a. It is important that, in contrast to a history (see Subsection 4.3.2), the *labels* of the identifiably preceding actions are 'listed'. Call the FSG $(G(\Gamma), F_{\mathrm{hist}})$ the history labelled agent normal form of Γ.

Reconsider, again, the extensive game Γ^{CH} from Subsection 4.3.1 and its simply labelled agent normal form $(G(\Gamma^{CH}), F^{CH})$ as described above. Let $(G(\Gamma^{CH}), F^{CH}_{\mathrm{hist}})$ be the corresponding history labelled agent normal form of Γ^{CH}, i.e., we have

$$\ell^*_{(i,0,0)a}(i, 0, 0, z_{1i}) = z_{1i} \tag{4.5a}$$

$$\ell^*_{(i,z_{11},z_{12})a}(i, z_{11}, z_{12}, z_{2i}) = z_{2i} \tag{4.5b}$$

$$\ell^*_{(i,0,0)(i',0,0,z_{1_{i'}})}(i,0,0,z_{1i}) = \text{no} \tag{4.5c}$$

$$\ell^*_{(i,0,0)(i',z'_{11},z'_{12},z'_{2_{i'}})}(i,0,0,z_{1i}) = \text{no} \tag{4.5d}$$

$$\ell^*_{(i,z_{11},z_{12})(i',0,0,z'_{1_{i'}})}(i,z_{11},z_{12},z_{2i}) = \begin{cases} z'_{1i'} & : z'_{1i'} = z_{1i'} \\ \text{no} & : z'_{1i'} \neq z_{1i'} \end{cases} \tag{4.5e}$$

$$\ell^*_{(i,z_{11},z_{12})(i',z'_{11},z'_{12},z'_{2_{i'}})}(i,z_{11},z_{12},z_{2i}) = \text{no} \tag{4.5f}$$

for all $(i,0,0,z_{1i})$, (i,z_{11},z_{12},z_{2i}), $(i',0,0,z'_{1_{i'}})$, $(i',z'_{11},z'_{12},z'_{2_{i'}}) \in A$. Justification and interpretation are very similar to Example 4.3.1. But the actions in the histories have to be replaced by their labels.

The history labelling is illustrated for some selected information sets in Figure 4.6 where $i = 1, 2$. Subfigure (a) shows the frame for the first round: All actions are labelled by their distinguishing numbers with regard to the attribute a and by the label no with regard to all of the history related attributes—the players cannot identify any preceding actions in the first round. This way, in a sense, all first-round actions remain symmetric. In Subfigure (b), the labelling of the subgame reached after coordination at the players' first actions in the first round is given. Now, the history attributes related to the preceding first-round actions—$(1,0,0,1)$ and $(2,0,0,1)$—transfer their distinguishing numbers—twice 1—to the actions as labels. By Corollary 2.5.1 (p. 44), neither the second nor the third action can be symmetric to first one. Finally, Subfigure (c) presents the labelling of the subgame after first-round discoordination at the first player's second action and the second player's third action. Again, by Corollary 2.5.1, neither the second nor the third action can be symmetric to first one.

These observations suggest that Crawford & Haller's rule can be supported in the fully labelled agent normal form of Γ. This is expressed with the following theorem.

Theorem 4.3.3.
In $(G(\Gamma^{CH}), F^{CH}_{\text{hist}})$, the unique payoff-dominant symmetry invariant equilibrium b^* is characterized by

1. $b^*_{(1,0,0)}(i,0,0,z_{1i}) = \frac{1}{3}$,
2. $b^*_{(1,z_{11},z_{12})}(i,z_{11},z_{12},z_{2i}) = 1$ *if $z_{11} = z_{12} = z_{2i}$,*
3. $b^*_{(1,z_{11},z_{12})}(i,z_{11},z_{12},z_{2i}) = 0$ *if $z_{11} = z_{12} \neq z_{2i}$,*
4. $b^*_{(1,z_{11},z_{12})}(i,z_{11},z_{12},z_{2i}) = 1$ *if $z_{11} \neq z_{12}$, $z_{2i} \neq z_{11}$, and $z_{2i} \neq z_{12}$, and*
5. $b^*_{(1,z_{11},z_{12})}(i,z_{11},z_{12},z_{2i}) = 0$ *if $z_{11} \neq z_{12}$ and $z_{2i} = z_{11}$ or $z_{2i} = z_{12}$*

for all $(i,0,0,z_{1i})$, $(i,z_{11},z_{12},z_{2i}) \in A$.

Proof. Let the bijections $\nu : H \to H$, $r_h : A_h \to A_{\pi(h)}$ for all $h \in H$, $\mu : \mathfrak{A}^* \to \mathfrak{A}^*$, and $\tau : \Lambda^* \to \Lambda^*$ be induced by the permutation q of $\{1,2,3\}$ and the permutations π_1, π_2 of $\{1,2\}$ as follows:

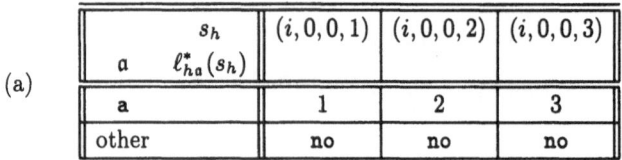

(a)

s_h $\mathfrak{a}$ $\ell^*_{h\mathfrak{a}}(s_h)$	$(i,0,0,1)$	$(i,0,0,2)$	$(i,0,0,3)$
a	1	2	3
other	**no**	**no**	**no**

(b)

s_h $\mathfrak{a}$ $\ell^*_{h\mathfrak{a}}(s_h)$	$(i,1,1,1)$	$(i,1,1,2)$	$(i,1,1,3)$
a	1	2	3
$(1,0,0,1)$	1	1	1
$(2,0,0,1)$	1	1	1
other	**no**	**no**	**no**

(c)

s_h $\mathfrak{a}$ $\ell^*_{h\mathfrak{a}}(s_h)$	$(i,2,3,1)$	$(i,2,3,2)$	$(i,2,3,3)$
a	1	2	3
$(1,0,0,2)$	2	2	2
$(2,0,0,3)$	3	3	3
other	**no**	**no**	**no**

Fig. 4.6. History labelling of selected information sets

$$\nu(i,0,0) = (\pi_1(i),0,0) \tag{4.6}$$
$$\nu(i,z_{11},z_{12}) = (\pi_2(i),q(z_{1\pi_1(1)}),q(z_{1\pi_1(2)})) \tag{4.7}$$

for all $(i,0,0),(i,z_{11},z_{12}) \in H$,

$$r_{(i,0,0)}(i,0,0,z_{1i}) = (\pi_1(i),0,0,q(z_{1i})) \tag{4.8}$$
$$r_{(i,z_{11},z_{12})}(i,z_{11},z_{12},z_{2i}) = (\pi_2(i),q(z_{1\pi_1(1)}),q(z_{1\pi_1(2)}),q(z_{2i})) \tag{4.9}$$

for all $(i,0,0,z_{1i}),(i,z_{11},z_{12},z_{2i}) \in A$,

$$\mu(\mathbf{a}) = \mathbf{a} \tag{4.10}$$
$$\mu(i,0,0,z_{1i}) = r_{(i,0,0)}(i,0,0,z_{1i}) \tag{4.11}$$
$$\mu(i,z_{11},z_{12},z_{2i}) = r_{(i,z_{11},z_{12})}(i,z_{11},z_{12},z_{2i}) \tag{4.12}$$

for all $(i,0,0,z_{1i}),(i,z_{11},z_{12},z_{2i}) \in A \subset \mathfrak{A}^*$,

$$\tau(\mathbf{no}) = \mathbf{no} \tag{4.13}$$
$$\tau(z) = q(z) \tag{4.14}$$

for all $z \in \{1,2,3\}$.

By (4.8) and (4.9), all actions are mapped in the 'same way' for both players with regard to their distinguishing numbers. In view of (4.1), condition

(2.5) of Definition 2.3.6 is satisfied by $(\nu, (r_h)_{h \in H})$. Hence, $(\nu, (r_h)_{h \in H})$ is symmetry of $G(\Gamma^{CH})$. In addition, the pair (μ, τ) together with $(\nu, (r_h)_{h \in H})$ meets condition (2.15) of Definition 2.5.4, i.e., $((\nu, (r_h)_{h \in H}), (\mu, \tau))$ is a symmetry of $(G(\Gamma^{CH}), F^{CH}_{hist})$. This can be seen from the following chains of equations which hold for all $(i, 0, 0, z_{1i}), (i, z_{11}, z_{12}, z_{2i}), (i', 0, 0, z'_{1i'}), (i', z'_{11}, z'_{12}, z'_{1i'}) \in A$:

We have

$$\ell^*_{\nu(i,0,0)\mu(a)}\big(r_{(i,0,0)}(i, 0, 0, z_{1i})\big) = \ell^*_{(\pi_1(i),0,0)a}\big(\pi_1(i), 0, 0, q(z_{1i})\big)$$
$$= q(z_{1i})$$
$$= \tau(z_{1i})$$
$$= \tau\big(\ell^*_{(i,0,0)a}(i, 0, 0, z_{1i})\big)$$

where the single equations follow from (4.6), (4.10), (4.8); (4.5a); (4.14); and (4.5a) respectively. Further we have

$$\ell^*_{\nu(i,z_{11},z_{12})\mu(a)}\big(r_{(i,z_{11},z_{12})}(i, z_{11}, z_{12}, z_{2i})\big) = \ldots$$
$$= \ell^*_{(\pi_2(i),q(z_{1\pi_1(1)}),q(z_{1\pi_1(2)}))a}\big(\pi_2(i), q(z_{1\pi_1(1)}), q(z_{1\pi_1(2)}), q(z_{1i})\big)$$
$$= q(z_{2i})$$
$$= \tau(z_{2i})$$
$$= \tau\big(\ell^*_{(i,z_{11},z_{12})a}(i, z_{11}, z_{12}, z_{2i})\big)$$

where the single equations follow from (4.7), (4.10), (4.9); (4.5b); (4.14); and (4.5b) respectively.

We also have

$$\ell^*_{\nu(i,0,0)\mu(i',0,0,z'_{1i'})}\big(r_{(i,0,0)}(i, 0, 0, z_{1i})\big) = \ldots$$
$$= \ell^*_{(\pi_1(i),0,0)(\pi_1(i'),0,0,q(z'_{1i'}))}\big(\pi_1(i), 0, 0, q(z_{1i})\big)$$
$$= \mathbf{no}$$
$$= \tau(\mathbf{no})$$
$$= \tau\big(\ell^*_{(i,0,0)(i',0,0,z'_{1i'})}(i, 0, 0, z_{1i})\big)$$

where the single equations follow from (4.6), (4.11), (4.8); (4.5c); (4.13); and (4.5c) respectively. Further, we have

$$\ell^*_{\nu(i,0,0)\mu(i',z'_{11},z'_{12},z'_{2i'})}\big(r_{(i,0,0)}(i, 0, 0, z_{1i})\big) = \ldots$$
$$= \ell^*_{(\pi_1(i),0,0)(\pi_2(i'),q(z'_{1\pi_1(1)}),q(z'_{1\pi_1(2)}),q(z'_{2i'}))}\big(\pi_1(i), 0, 0, q(z_{1i})\big)$$
$$= \mathbf{no}$$
$$= \tau(\mathbf{no})$$
$$= \tau\big(\ell^*_{(i,0,0)(i',z'_{11},z'_{12},z'_{2i'})}(i, 0, 0, z_{1i})\big)$$

where the single equations follow from (4.6), (4.12), (4.8); (4.5d); (4.13); and (4.5d) respectively.

Since π_1 is just a permutation of $\{1,2\}$, $\pi_1(\pi_1(i')) = i'$ hold for all i'. Therefore, $q(z'_{1i'}) = q(z_{1\pi_1(\pi_1(i'))})$ is equivalent to $q(z'_{1i'}) = q(z_{1i'})$, and even to $z'_{1i'} = z_{1i'}$ as q is bijective. For $q(z'_{1i'}) = q(z_{1\pi_1(\pi_1(i'))})$ or, equivalently, $z'_{1i'} = z_{1i'}$, we have

$$\ell^*_{\nu(i,z_{11},z_{12})\mu(i',0,0,z'_{1_*})}(r_{(i,z_{11},z_{12})}(i,z_{11},z_{12},z_{2i})) = \ldots$$
$$= \ell^*_{(\pi_2(i),q(z_{1\pi_1(1)}),q(z_{1\pi_1(2)}))(\pi_1(i'),0,0,q(z'_{1_*}))}(\pi_2(i),q(z_{1\pi_1(1)}),q(z_{1\pi_1(2)}),q(z_{2i}))$$
$$= q(z'_{1i'})$$
$$= \tau(z'_{1i'})$$
$$= \tau(\ell^*_{(i,z_{11},z_{12})(i',0,0,z'_{1_*})}(i,z_{11},z_{12},z_{2i}))$$

where the single equations follow from (4.7), (4.11), (4.9); (4.5e); (4.14); and (4.5e) respectively. For $q(z'_{1i'}) \neq q(z_{1\pi(\pi(i'))})$ or, equivalently, $z'_{1i'} \neq z_{1i'}$, we have

$$\ell^*_{\nu(i,z_{11},z_{12})\mu(i',0,0,z'_{1_*})}(r_{(i,z_{11},z_{12})}(i,z_{11},z_{12},z_{2i})) = \ldots$$
$$= \ell^*_{(\pi_2(i),q(z_{1\pi_1(1)}),q(z_{1\pi_1(2)}))(\pi_1(i'),0,0,q(z'_{1_*}))}(\pi_2(i),q(z_{1\pi_1(1)}),q(z_{1\pi_1(2)}),q(z_{2i}))$$
$$= \mathbf{no}$$
$$= \tau(\mathbf{no})$$
$$= \tau(\ell^*_{(i,z_{11},z_{12})(i',0,0,z'_{1_*})}(i,z_{11},z_{12},z_{2i}))$$

where the single equations follow from (4.7), (4.11), (4.9); (4.5e); (4.13); and (4.5e) respectively. Finally, we have

$$\ell^*_{\nu(i,z_{11},z_{12})\mu(i',z'_{11},z'_{12},z'_{2_*})}(r_{(i,z_{11},z_{12})}(i,z_{11},z_{12},z_{2i})) = \ldots$$
$$= \ell^*_{(\pi_2(i),q(z_{1\pi_1(1)}),q(z_{1\pi_1(2)}))(\pi_2(i'),q(z'_{1\pi_1(1)}),q(z'_{1\pi_1(2)}),q(z'_{2_*}))}(\pi_2(i),\ldots$$
$$\ldots q(z_{1\pi_1(1)}),q(z_{1\pi_1(2)}),q(z_{2i}))$$
$$= \mathbf{no}$$
$$= \tau(\mathbf{no})$$
$$= \tau(\ell^*_{(i,z_{11},z_{12})(i',z'_{11},z'_{12},z'_{2_*})}(i,z_{11},z_{12},z_{2i}))$$

where the single equations follow from (4.7), (4.12), (4.9); (4.5f); (4.13); and (4.5f) respectively.

Obviously, the symmetry of $(G(\Gamma^{CH}), F_{\text{hist}}^{CH})$ induced by q and π with $q(1) = 2$, $q(2) = 3$, $q(3) = 1$ and $\pi_1 = \pi_2 = \text{id}$ establishes the symmetry of all of a player's first-round actions. This proves the first claim.

Consider now the symmetry $((\nu, (r_h)_{h \in H}), (\mu, \tau))$ induced by $q(1) = 1$, $q(2) = 3$, $q(3) = 2$ and $\pi_1 = \pi_2 = \mathrm{id}$. By (4.9), then, $r_{(i,1,1)}(i, 1, 1, 2) = (i, 1, 1, 2)$ for both i. I.e., the second and third action after coordination at the players' first actions in the first round—at the information sets $(1, 1, 1)$ and $(2, 1, 1)$—are symmetric. In view of the labelling (4.5) as illustrated by Figure 4.6 (b) and Corollary 2.5.1, as mentioned above, neither a player's second nor his third action can be symmetric to the first one. Thus, the second and the third action have to get the same probabilities in a symmetry variant behavior strategy combination. Payoff dominance then requires to put probability 1 on the first actions and to put probability 0 on the other actions. Analogous arguments can be applied to first-round coordination at the other actions. This proves the second and the third claim.

The symmetry $((\nu, (r_h)_{h \in H}), (\mu, \tau))$ induced by $q(1) = 1$, $q(2) = 3$, $q(3) = 2$, $\pi_1(1) = 2$, $\pi_1(2) = 1$, and $\pi_2 = \mathrm{id}$ gives $r_{(i,2,3)}(i, 2, 3, 2) = (i, 2, 3, 3)$ for both i. I.e., a player's second and his third action are symmetric after first-round discoordination at player 1's second and player 2's third action, i.e., at the information sets $(1, 2, 3)$ and $(2, 2, 3)$. Again, in view of the labelling shown in Figure 4.6 (c) and Corollary 2.5.1, neither a player's second nor his third action can be symmetric to his first action after first-round discoordination at player 1's second and player 2's third action, i.e., at the information sets $(1, 2, 3)$ and $(2, 2, 3)$. Payoff dominance then requires to put probability 1 on the first actions and to put probability 0 on the other actions. Analogous arguments can be applied to first-round discoordination at the other actions. This proves the remaining claims. $\square$

In making up a history labelled agent normal form, some abuse is made of the frame: Remember that frames were introduced in order to formalize the players' *non-strategic* descriptions of a game. But here, frames are used to represent the sequence of moves to some extent which clearly is a strategic feature of an extensive game. In addition, the history frames of agent normal forms are complex and rather difficult to deal with. So it seems to be both more natural and less cumbersome to apply the framing to extensive games in which the order of moves already is incorporated directly. This is done in the next section.

4.4 Framed Extensive Games (FEGs)

Framed extensive games (FEGs) are defined by analogy with FSGs that have been introduced and motivated in detail in Chapter 2.

4.4.1 Frames

Definition 4.4.1 (Frames of extensive games).
A frame Φ of an extensive game $\Gamma = (T, \lhd, I, P, H, A, p, u)$ is a triple $(\mathfrak{A}, \Lambda,$

$(\ell_\mathfrak{a})_{\mathfrak{a} \in \mathfrak{A}}$), where $\mathfrak{A}$ denotes a non-empty set of attributes $\mathfrak{a}$, Λ a non-empty set of labels λ, and $\ell_\mathfrak{a}$ the label function $A \to \Lambda$ for each attribute $\mathfrak{a}$.

The pair $(\mathfrak{A}, \Lambda)$ is called the language of the frame Φ and $(\ell_\mathfrak{a})_{\mathfrak{a} \in \mathfrak{A}}$ its labelling. The attributes $\mathfrak{a}$ represent properties of the actions (e.g., color or shape of an object to choose), and the labels λ characterize the individual actions with respect to one attribute. A frame $\Phi = (\mathfrak{A}, \Lambda, (\ell_\mathfrak{a})_{\mathfrak{a} \in \mathfrak{A}})$ is called finite if both $\mathfrak{A}$ and Λ are finite, and it is called simple or a simple labelling if the attribute set contains just one attribute—$|\mathfrak{A}| = 1$.

Example 4.4.1.
Consider the extensive game Γ^{CH} from Section 4.3.1. The following frame assigns the distinguishing numbers to any action as the frame F^{CH} of Γ^{CH}'s agent normal form does (see Section 4.3.4). We set $\mathfrak{A} = \{\mathfrak{a}\}$, $\Lambda = \{1, 2, 3\}$, and $\ell_\mathfrak{a}(i, z_1, z_2, z) = z$ for all $(i, z_1, z_2, z) \in A$.

A pair (Γ, Φ) is called *framed extensive game* (FEG). A FEG (Γ, Φ) is called finite if both Γ and Φ are finite; $\mathcal{F}^e$ denotes the class of finite FEGs. Equilibria of FEGs are the equilibria of the underlying extensive games. The following extension of solution concepts for extensive games makes explicit the idea that different framings of a given extensive game may result in different solutions:

Definition 4.4.2 (Solution concepts for FEGs).
A solution concept L for $\mathcal{F}^e$ assigns a set of behavior-strategy combinations $L(\Gamma, \Phi) \subseteq B$ to every FEG $(\Gamma, \Phi) \in \mathcal{F}^e$.

4.4.2 Weak Isomorphisms of FEGs and Language Invariance

The motivation for the following definition is the same as for strategic games (see Subsections 2.5.1 and 2.5.2). By analogy with FSGs, weak isomorphisms of extensive games are extended to the class of FEGs:

Definition 4.4.3 (Weak isomorphisms of FEGs).
Let (Γ, Φ) and $(\bar{\Gamma}, \bar{\Phi})$ be FEGs. A weak isomorphism from (Γ, Φ) onto $(\bar{\Gamma}, \bar{\Phi})$ is a weak isomorphism r from Γ to $\bar{\Gamma}$ together with a pair of bijections $\mathbf{t} = (\mu, \tau)$, $\mu : \mathfrak{A} \to \bar{\mathfrak{A}}$ and $\tau : \Lambda \to \bar{\Lambda}$, such that

$$\tau(\ell_\mathfrak{a}(a)) = \bar{\ell}_{\mu(\mathfrak{a})}(r(a)) \tag{4.15}$$

holds for all $a \in A$ and $\mathfrak{a} \in \mathfrak{A}$.

Obviously, a system of identity mappings is a weak isomorphism from a FEG to itself. Since all mappings involved are bijective, inverses and compositions of weak isomorphisms are also weak isomorphisms. Thus, weak isomorphisms constitute an equivalence relation on $\mathcal{F}^e$. Two FEGs are called weakly isomorphic if they can be mapped onto each other by a weak isomorphism. The next definition is analogously motivated as the very similar definition for FSGs in Subsection 2.5.2.

Definition 4.4.4 (Language invariance).
A solution concept L for $\mathcal{F}^e$ is invariant with respect to weak isomorphisms (language invariant) if for every isomorphism $(r, \mathbf{t})$ from $(\Gamma, \Phi) \in \mathcal{F}^e$ to $(\bar{\Gamma}, \bar{\Phi}) \in \mathcal{F}^e$ we have $\mathbf{r}(L(\Gamma, \Phi)) = L(\bar{\Gamma}, \bar{\Phi})$ for $\mathbf{r}$ given by r via (3.15).

4.4.3 Weak Symmetries of FEGs

In FEGs, weak symmetry invariance and the weak symmetry of actions are defined as in extensive games except for that weak symmetries of the underlying extensive game that are part of weak symmetries of FEGs are considered only. Since by definition a smaller set of extensive-game weak symmetries is considered, Theorem 3.3.3 (p. 74) also holds for FEGs.

Theorem 4.4.1.
Any FEG has a weakly symmetry invariant perfect equilibrium.

4.5 Modelling Dynamic Focal Points by FEGs

Now, using FEGs, we can formalize dynamic focal points in a rather natural way. Firstly, the decision problem is represented by an extensive game, and then, an appropriately chosen frame models the relatedness of actions as for example actions that represent choosing the same object. Finally, the symmetry invariance criterion together with payoff dominance selects from the equilibria of the resulting FEG. In a sense, a unique payoff-dominant weakly symmetry-invariant perfect equilibrium can be considered as the focal point or the focal point equilibrium of the game.

Using the examples from Section 4.1, we now demonstrate how weak symmetry invariance in FEGs can be used in order to support our intuitions about the dynamic focal points in these games.

4.5.1 Crawford & Haller's Rule

Consider the extensive game Γ^{CH} from Section 4.3.1. Let the frame Φ^{CH} of Γ^{CH} be equivalent to the simple frame F^{CH} of Γ^{CH}'s agent normal form from Section 4.3.4. I.e., the constituents of Φ^{CH} are such that $\mathfrak{A} = \{a\}$, $\Lambda = \{1, 2, 3\}$, and $\ell_a(i, z_1, z_2, z) = z$ for all $(i, z_1, z_2, z) \in A$. The player can identify second-round actions with the corresponding first-round actions.

Theorem 4.5.1.
In (Γ^{CH}, Φ^{CH}), the unique payoff-dominant weakly symmetry-invariant equilibrium b^* is characterized by

1. $b^*_{(i,0,0)}(i, 0, 0, z_{1i}) = \frac{1}{3}$,
2. $b^*_{(i,z_{11},z_{12})}(i, z_{11}, z_{12}, z_{2i}) = 1$ if $z_{11} = z_{12} = z_{2i}$,

3. $b^*_{(i,z_{11},z_{12})}(i,z_{11},z_{12},z_{2i}) = 0$ if $z_{11} = z_{12} \neq z_{2i}$,

4. $b^*_{(i,z_{11},z_{12})}(i,z_{11},z_{12},z_{2i}) = 1$ if $z_{11} \neq z_{12}$, $z_{2i} \neq z_{11}$, and $z_{2i} \neq z_{11}$, and

5. $b^*_{(i,z_{11},z_{12})}(i,z_{11},z_{12},z_{2i}) = 0$ if $z_{11} \neq z_{12}$ and $z_{2i} = z_{11}$ or $z_{2i} = z_{11}$

for all $(i,0,0,z_{1i}), (i,z_{11},z_{12},z_{2i}) \in A$.

I.e., in the FEG framework, we can support Crawford & Haller's rule. But, of course, the restriction to two rounds circumvents the ambiguities discussed in Section 4.2.

Proof. Let the bijection $r : A \to A$ be defined by the bijections $\pi : I \to I$ and $q : \{1,2,3\} \to \{1,2,3\}$ as follows: $r(i,0,0,z_{1i}) = (\pi(i),0,0,q(z_{1i}))$ and $r(i,z_{11},z_{12},z_{2i}) = (\pi(i),q(z_{1\pi(1)}),q(z_{1\pi(2)}),q(z_{2i}))$ for all $(i,0,0,z_{1i}), (i,z_{11}, z_{12},z_{2i}) \in A$. By **ISA**, this induces the bijection $\nu : H \to H$ with $r(i,0,0) = (\pi(i),0,0)$ and $r(i,z_{11},z_{12},z_{2i}) = (\pi(i),q(z_{1\pi(1)}),q(z_{1\pi(2)}))$ for all $(i,0,0),(i, z_{11},z_{12}) \in H$, and by **PTH**, this induces the bijection $\theta : Z \to Z$ with $\theta(z_{11}, z_{12},z_{21},z_{22}) = (q(z_{1\pi(1)}),q(z_{1\pi(2)}),q(z_{2\pi(1)}),q(z_{2(2)}))$. The latter can be seen from the following equations

$$r(\psi(z_{11},z_{12},z_{21},z_{22})) = \{r(1,0,0,z_{11}),r(2,0,0,z_{12}),r(1,z_{11},z_{12},z_{21}),\dots$$
$$\dots r(2,z_{11},z_{12},z_{22})\}$$
$$= \{(\pi(1),0,0,q(z_{11})),(\pi(2),0,0,q(z_{12})),\dots$$
$$\dots (\pi(1),q(z_{1\pi(1)}),q(z_{1\pi(2)}),q(z_{2\pi(1)})),\dots$$
$$\dots (\pi(2),q(z_{1\pi(1)}),q(z_{1\pi(2)}),q(z_{2\pi(2)}))\}$$
$$= \psi(q(z_{1\pi(1)}),q(z_{1\pi(2)}),q(z_{2\pi(1)}),q(z_{2\pi(2)})))$$

where the single equations are implied by Definition 4.3.1; the setting of r above; and Definition 4.3.1 respectively. Since the action-indices (last components) are mapped in the same way for both players, condition **PY** is also satisfied. Thus, r is a weak symmetry of Γ^{CH}. In addition, we have

$$\ell_{\mu(a)}(r(i,0,0,z_{1i})) = \ell_a(\pi(i),0,0,q(z_{1i}))$$
$$= q(z_{1i})$$
$$= \tau(\ell_a(i,0,0,z_{1i}))$$

and

$$\ell_{\mu(a)}(r(i,z_{11},z_{12},z_{2i})) = \ell_a(\pi(i),q(z_{1\pi(1)}),q(z_{1\pi(2)}),q(z_{2i}))$$
$$= q(z_{2i})$$
$$= \tau(\ell_a(i,z_{11},z_{12},z_{2i})))$$

where the single equations of each chain follow from the setting of r and μ above; the setting of Φ^{CH}; and the setting of τ above, respectively. Hence, $(r,(\mu,\tau))$ satisfies (4.15), i.e., $(r,(\mu,\tau))$ is a weak symmetry of (Γ^{CH},Φ^{CH}).

Consider the weak symmetry $(r, (\mu, \tau))$ of (Γ^{CH}, Φ^{CH}) defined by $\pi = \mathrm{id}$ and $q(1) = 2$, $q(2) = 3$, and $q(3) = 1$. Obviously, this symmetry establishes the symmetry of the players' first-round actions. This proves the first claim.

Consider now the weak symmetry $(r, (\mu, \tau))$ of (Γ^{CH}, Φ^{CH}) defined by $\pi = \mathrm{id}$ and $q(1) = 1$, $q(2) = 3$, and $q(3) = 2$. This implies $r(i, 1, 1, 2) = (i, 1, 1, 3)$ for both i, i.e., the players' second and third actions are symmetric after coordination at their first actions in the first round. However, neither the second nor the third action of this information set can be symmetric to the first one for any of the players. Suppose on the contrary, the actions $(1, 1, 1, 1)$ and $(1, 1, 1, 2)$ were symmetric, i.e., there were weak symmetry $(r, (\mu, \tau))$ of (Γ^{CH}, Φ^{CH}) such that $r(1, 1, 1, 1) = (1, 1, 1, 2)$. By definition of Γ^{CH}, we had $(1, 1, 1, 1) \in a(\psi(1, 1, 1, 1))$. Then, by assumption and by **PTH**, we had $(1, 1, 1, 2) \in r(a(\psi(1, 1, 1, 1))) = a(\psi(\theta(1, 1, 1, 1)))$. The definition of Γ^{CH} and **PTH**, again, imply $(1, 0, 0, 1) \in a(\psi(\theta(1, 1, 1, 1))) = r(a(\psi(1, 1, 1, 1)))$. Hence, by definition of Φ^{CH} and (4.15), we have $\{1, 2\} \subseteq \ell_a(r(a(\psi(1, 1, 1, 1)))) = \tau(\ell_a(a(\psi(1, 1, 1, 1)))) = \tau(\{1\})$. Clearly, this contradicts τ's being bijective. Finally, payoff dominance selects the players' first actions after coordination at their first actions in the first round. Analogously, one can show that this also holds for first-round coordination at the other actions. This proves the second and the third claim.

Consider now discoordination at the actions $(1, 0, 0, 2)$ and $(2, 0, 0, 3)$ in the first round. A bit surprisingly, $(i, 2, 3, 2)$ and $(i, 2, 3, 3)$ are not symmetric. But the intuition behind this is clear: In the case of non-coordination in the first round, a player can distinguish the action(-index) he actually took in the first round and the first-round action that would have led to coordination given the other player's choice. Formally: Suppose on the contrary there were a symmetry r of (Γ^{CH}, Φ^{CH}) such that $r(i, 2, 3, 2) = (i, 2, 3, 3)$. By (4.15), we had $\tau(2) = 3$, and by **PL**, we had $\pi = \mathrm{id}$. Then, **PTH** implies $r(i, 0, 0, 2) = (i, 0, 0, 2)$, and by (4.15) we had $\tau(2) = 2$—a contradiction. Instead one can show that the symmetry $(r, (\mu, \tau))$ of (Γ^{CH}, Φ^{CH}) defined by $\pi(1) = 2$, $\pi(2) = 1$, $q(1) = 1$, $q(2) = 3$, and $q(3) = 2$ establishes the symmetry of the actions $(1, 2, 3, 2)$ and $(2, 2, 3, 3)$, and the symmetry of $(1, 2, 3, 3)$ and $(2, 2, 3, 2)$.

In addition, it is easy to check that neither $(i, 2, 3, 2)$ nor $(i, 2, 3, 3)$ can be symmetric to $(i, 2, 3, 1)$. Hence, in view of (4.1), payoff dominance requires each player to take his first action. The other cases of discoordination in the first round can be treated analogously. This proves the last two claims. $\square$

4.5.2 Conscious Production of a Description

Reconsider the stage game from Section 4.1 in Figure 4.1. The frame given in Figure 4.7 can be used to connect actions that refer to taking the same object in the different stages of the repeated game. In view of Theorem 4.5.1, the unique payoff-dominant symmetry-invariant perfect equilibrium of the

a $\quad$ $\ell_{ia}(s_{ij})$ $\qquad s_{ij}$	s_{i1}	s_{i2}	s_{i3}	s_{i4}
a	1	2	3	4

Fig. 4.7. Simple frame of a stage game

a $\quad$ $\ell_{ha}^{*}(a)$ $\qquad a$	$\{x\}$	$\{z\}$
a	λ	$\lambda^{'}$
$\{x\}$	no	λ
$\{z\}$	no	no

Fig. 4.8. A history labelling

straightforward FEG representation of the two-stage game is (a) randomizing over the last three actions at the first stage, (b) repeating the coordinating stage game actions of the first stage, and (c) in case of discoordination, switching to that stage game action that a player did not use at the previous stage and that had not led to coordination given the other player's first-round action. Again, this is also the intuitive solution.

4.5.3 History Labelled Agent Normal Forms

Theorem 4.3.2 suggests that the weak symmetry of actions in simply labelled extensive games is transferred to the symmetry of actions in their history labelled agent normal forms in non-pathological cases. The following example, involving a pathological game, shows that this does not hold generally.

Example 4.5.1.
Reconsider the extensive game Γ from Example 4.3.2 (p. 98) together with the simple frame Φ with $\mathfrak{A} = \{a\}$, $\Lambda = \{\lambda, \lambda^{'}\}$, and $\ell_{1a}(\{x\}) = \lambda$, $\ell_{1a}(\{z\}) = \lambda^{'}$. Clearly, the actions $\{x\}$ and $\{z\}$ are weakly symmetric in the simply labelled extensive game (Γ, Φ). Consider now the history labelled agent normal form $(G(\Gamma), F_{\text{hist}})$ of (Γ, Φ) which has the labelling given in Figure 4.8. By Corollary 2.5.1 (p. 44), $\{x\}$ and $\{z\}$ cannot be symmetric.

Excluding pathological cases, indeed, the conjecture above can be corroborated.

Theorem 4.5.2.
Let (Γ, Φ) be some simply labelled extensive game where Γ is non-pathological. Let further be $(G(\Gamma), F_{\text{hist}}(\Phi))$ the history labelled agent normal form as defined in Section 4.3.4 based on the frame $F(\Phi)$, where $F(\Phi)$'s and Φ's attribute set and label set are identical, and the label functions are related in the following way:

$$\ell_{ha}(a) := \ell_a(a) \qquad (4.16)$$

for all $h \in H \backslash H_{i_0}$ and $a \in A_h$. Then, weakly symmetric actions in (Γ, Φ) are also symmetric in $(G(\Gamma), F_{\text{hist}}(\Phi))$.

Proof. Let (Γ, Φ) be a non-pathological simply labelled extensive game, and let $(r, (\mu, \tau))$ be a weak symmetry of (Γ, Φ). Let further $\pi : I \to I, \nu : H \to H$, and $\theta : Z \to Z$ be the bijections induced by r according to Definition 3.3.1 (p. 67). Consider the restriction ν^* of ν to the domain $H \backslash H_{i_0}$ which, by **CPL**, is a bijection of $H \backslash H_{i_0}$ onto itself. Consider further the bijections $(r_h^*)_{h \in H \backslash H_{i_0}}$, $r_h^* : A_h \to A_{\nu^*(h)}$ satisfying $r_h^*(a) = r(a)$ for all $h \in H \backslash H_{i_0}$ and $a \in A_h$; the bijection $\mu^* : \mathfrak{A}^* \to \mathfrak{A}^*$ satisfying $\mu^*(\mathbf{a}) = \mathbf{a}$ and $\mu^*(a) = r(a)$ for all $a \in A \subset \mathfrak{A}^*$; and the bijection $\tau^* : \Lambda^* \to \Lambda^*$ satisfying $\tau^*(\mathbf{no}) = \mathbf{no}$ and $\tau^*(\lambda) = \tau(\lambda)$ for all $\lambda \in \Lambda \subset \Lambda^*$. Remains to show that $((\nu^*, (r_h^*)_{h \in H \backslash H_{i_0}}), (\mu^*, \tau^*))$ is a symmetry of $(G(\Gamma), F_{\text{hist}}(\Phi))$.

Using the argumentation in the proof of Theorem 3.3.2 (p. 74), one can show that $(\nu^*, (r_h^*)_{h \in H \backslash H_{i_0}})$ also is a symmetry of $G(\Gamma)$. In addition, we have $\ell_{\nu^*(h)\mu^*(\mathbf{a})}^*(r_h(a)) = \tau^*(\ell_{ha}^*(a))$ for all $h \in H \backslash H_{i_0}$ and $a \in A_h$ which can be seen from

$$\ell_{\nu^*(h)\mu^*(\mathbf{a})}^*(r_h(a)) = \ell_{\nu^*(h)\mu^*(\mathbf{a})}^*(r_h(a))$$
$$= \ell_{\nu(h)\mathbf{a}}(r(a))$$
$$= \tau(\ell_{ha}(a))$$
$$= \tau^*(\ell_{ha}^*(a))$$

where the single equations follow from (4.4); the settings above for ν^*, μ^*, and r_h; (4.15); and the setting above for τ^* respectively. Using the argumentation from the proof of Theorem 4.3.2 (p. 99), we have $a' \in \text{hist}(h)$ if and only if $r(a') \in \text{hist}(\nu(h))$ for all $a' \in A$ and $h \in H \backslash H_{i_0}$. In view of the settings above for ν^* and μ^*, $a' \in \text{hist}(h)$ also is equivalent to $\mu^*(a') \in \text{hist}(\nu^*(h))$. So for $a' \in \text{hist}(h)$ or equivalently $\mu^*(a') \in \text{hist}(\nu^*(h))$ we have

$$\ell_{\nu^*(h)\mu^*(a')}^*(r_h(a)) = \ell_{\nu(h')\mathbf{a}}(\mu^*(a'))$$
$$= \ell_{\nu(h')\mathbf{a}}(r(a'))$$
$$= \tau(\ell_{h'\mathbf{a}}(a'))$$
$$= \tau^*(\ell_{ha'}^*(a))$$

for all $h \in H \backslash H_{i_0}$, $a \in A_h$, $h' \in H$ and $a' \in A \subset \mathfrak{A}^*$ where the single equations follow from (4.4); the setting above for μ^*; (4.15); and the setting above for τ^* and (4.4) respectively. For $a' \notin \text{hist}(h)$ or equivalently $\mu^*(a') \notin \text{hist}(\nu^*(h))$, we have

$$\ell_{\nu^*(h)\mu^*(a')}^*(r_h(a)) = \mathbf{no}$$
$$= \tau^*(\mathbf{no})$$
$$= \tau^*(\ell_{ha'}^*(a))$$

for all $h \in H \backslash H_{i_0}$, $a \in A_h$, $h' \in H$ and $a' \in A \subset \mathfrak{A}^*$ where the single equations follow from (4.4); and the setting above for τ^*; and (4.4) respectively. Hence, (2.15) is satisfied by $((\nu^*, (r_h^*)_{h \in H \backslash H_{i_0}}), (\mu^*, \tau^*))$. $\square$

As the following example shows, the converse does not hold in general.

Example 4.5.2.
Reconsider the game (Γ^{CH}, Φ^{CH}) from Section 4.5.1. In the proof of Theorem 4.5.1, it was shown that the actions $(1, 2, 3, 2)$ and $(1, 2, 3, 3)$ are not symmetric. But, as the proof of Theorem 4.3.3 reveals, $(1, 2, 3, 2)$ and $(1, 2, 3, 3)$ are symmetric in the history labeled agent normal form $(G(\Gamma^{CH}), F_{\text{hist}}(\Phi^{CH}))$ of (Γ^{CH}, Φ^{CH}).

5 Conclusion

And out of the ground the LORD God formed every beast of
the field, and every fowl in the air; and brought *them* unto
Adam to see what he would call them: and whatsoever Adam
called every living creature, that *was* the name thereof.
And Adam gave names to all cattle, and to the fowl in the
air, and to every beast of the field; ... (Genesis 2.19–20)

5.1 Static Focal Points in Framed Strategic Games

Strategic games do not represent the players' description of the strategies—
the strategy-indices and player-indices are the *game theorist's* labels for
strategies and players. Hence, there is much arbitrariness in choosing them—
in general, two game theorists will use different labels. Therefore, the solutions
of strategic games should not be affected by the labels. This idea is formalized with Harsanyi & Selten's (1988) requirement of invariance with respect
to isomorphisms on the solutions of strategic games.

Since in pure coordination games all strategies are symmetric, invariance with respect to isomorphisms prescribes randomization. However, this
is highly inefficient—it renders the players with payoffs far lower than in the
case of coordination.

Observations in real life as well as in controlled experiments reveal that
players quite often can 'break' the symmetry of their strategies and coordinate
much more frequently than by randomization. Schelling (1960) introduces
the notion of focal points in order to explain this phenomenon—players often
find some symmetry-breaking clue in their description of a game. These focal
points are based on non-strategic features of games, namely, the players'
labelling of strategies and players. For that reason, focal points cannot be
explained within the framework of traditional game theory.

Only recently have attempts been made to formalize the idea of salience
that underlies static focal points. Bacharach (1991, 1993), Sugden (1995),

Janssen (1997), and Bacharach & Stahl (2000) formalize the players' descriptions of the strategies and, in many cases, derive focal points and salience in an intuitive way. Unfortunately, these approaches apply to special classes of games only and leave some subtle kinds of salience unexplained.

In Chapter 2, we introduce a general framework for the explanation of focal points as symmetry-breaking devices in strategic games—the FSG (framed strategic game) approach. This approach combines traditional game theory with the focal point theories above:

The focal point theories contribute a structure called *frame*. These frames represent the players' apprehension of the game in a way similar to Bacharach's (1991, 1993) VFT (variable frame theory) via multidimensional strategy labels. For this purpose, frames contain attributes, labels, and label functions. The attributes can be thought of as representing properties of the strategies (e.g., color or shape of an object to choose) as they are perceived by the players, and a label characterizes the individual strategies with respect to one property (e.g., the redness of an object). A strategy might be characterized by several attributes. The label functions connect the strategies with the attributes and labels—all strategies are assigned a label with respect to all of the attributes.

Traditional game theory contributes the *language invariance* requirement: Even in our theory, some arbitrariness remains. In making up FSGs, two game theorists formalizing the same decision problem are likely to use different expressions to denote the players' attributes and labels. However, the solution of a FSG should be independent of these arbitrary expressions as it should be independent of the game theorist's labels of players and strategies. For that reason, Harsanyi & Selten's (1988) strong requirement of invariance with respect to isomorphisms is extended to FSGs.

Then, focal points can be defined as unique payoff-dominant symmetry invariant equilibria of FSGs. Although the language invariance requirement that underlies symmetry invariance is very strong, in a very general manner, much of our intuition about focal points can be supported within the FSG framework. It is demonstrated that essential features of the other formalizations of focal points can be represented by the FSG approach.

5.2 Weak Isomorphisms of Extensive Games

Isomorphisms of extensive games can be viewed as a means to identify *structurally* similar extensive games and to identify corresponding structural elements of these games—players, information sets, actions, and nodes. While there is an established notion of isomorphisms of strategic games (Harsanyi & Selten 1988), there is no such notion for extensive games. Mainly, the reason for this gap seems to be the more complex nature of extensive games in connection with the resulting ambiguities in interpreting them.

The ambiguity that is most important for isomorphisms concerns the order of moves. One interpretation takes this order literally—the order of actions corresponds to the order in which the players move in reality. Another interpretation primarily regards the order of actions as a technical feature of extensive games that need not correspond to a real order of moving. While the latter interpretation is compatible with the traditional representation of strategic games as extensive games, the former is not. In the present context, compatibility means that symmetric strategies in a strategic game are symmetric if and only if the actions representing them are symmetric in any extensive game representation of that strategic game.

The strong isomorphisms introduced by Elmes & Reny (1994) and Peleg et al. (1999) rest upon bijections of the node sets which respect the order of moves in a very strong sense. This way, they fit with the literal interpretation of the order of moves. So it is not astonishing that these strong isomorphisms are incompatible with the traditional extensive representations of strategic games. In order to remedy this incompatibility, Peleg et al. (2000) introduce a new kind of extensive representation—*canonical extensive forms of game forms*. However, these representations are much more complex than the traditional ones.

The symmetries of extensive games introduced by Selten (1983), Oh (1995), and Casajus (1998) are based on bijections of the action partitions. They give rise to the weak isomorphisms suggested in Chapter 3. As it turns out, weak isomorphisms are compatible with the traditional extensive representations of strategic games. One the other hand, weak isomorphisms do not respect the order of moves in the strict sense. So they are related the loose interpretation of the order of moves in extensive games. This raises the question whether weak isomorphisms respect the order of moves in the essential extent?

The following criterion for this essential extent was suggested (see Definition 3.3.5, p. 78): A concept of isomorphisms respects the order of moves to the essential extent if sound solution concepts are invariant with respect to these isomorphisms, i.e., isomorphisms carry solutions of the kind under consideration into solutions of the same kind. This property is especially important with respect to solution concepts that explicitly refer to the sequential nature of moves as the concepts of subgame perfect equilibria and sequential equilibria.

In non-pathological cases, weak isomorphisms respect the order of moves to the essential extent. In addition, the symmetry invariance requirement based on weak isomorphisms is compatible with the considered equilibrium concepts: even symmetry invariant perfect equilibria do always exist. This justifies the use of weak isomorphisms by Selten (1983, 1988), Oh (1995), and Casajus (1998) as well as their use in Chapter 4.

5.3 Dynamic Focal Points in Framed Extensive Games

Symmetry-breaking dynamic focal points cannot be formalized in extensive games. It turns out that Oh's (1995) history approach does not refine weak isomorphisms substantially. Therefore, it also is unable to explain dynamic focal points.

Dynamic focal points can be explained within the FSG framework. But this requires modelling the sequence of moves—a strategic feature of an extensive games—by frames. I.e., some abuse is made of the frames that originally were introduced in order to formalize the players' *non-strategic* descriptions of a game. In addition, the required frames are complex and difficult to deal with.

For that reason, the FSG approach is transferred to extensive games— extensive games are extended by frames to *framed extensive games* (FEGs), and invariance with respect to weak isomorphisms is extended to a language invariance requirement. We demonstrate with some examples that dynamic focal points can be formalized as the unique payoff-dominant symmetry invariant equilibria of FEGs.

5.4 Perspective

In this book, we provide a general framework for incorporating the players' perceptions of games into game theory—the framed games approach. Within this framework, much of our intuition about focal points and salience can be supported.

To complement these findings, future research should be directed towards an experimental verification of some implications of our approach. In that context, it seems to be especially interesting to do experiments concerning the subtle salience which can be explained within the FSG approach (see Sections 2.2 and 2.5.6). Subtle salience requires both attributes, color and shape, to come to the mind of the players with the same intensity—a requirement that might be difficult to be ensured in an experimental set-up. Therefore, the other way round, adherence to subtle salience itself could be used in order to assess whether the attributes come to the mind with the same intensity, i.e., the players' choice of the subtly salient option is used as an indicator for this fact. By varying the brightness of objects in Example 2.2.1, one could explore which difference of brightness just outweighs the objects' shape as a coordinating device. Of course, this takes for granted *that* the players adhere to this subtle kind of salience.

Another issue of future research concerns the weak isomorphisms and the related weak symmetry invariance requirement. It seems to be worthwhile to explore the connections to applied game theory. Especially, behavioral norms in dynamic settings (e.g., De Francesco 1996) should be analyzed whether they can be interpreted as the (unique) payoff-dominant weakly symmetry invariant equilibria of the underlying games.

List of Symbols and Abbreviations

$\mathfrak{A}$	attribute set
$\mathfrak{a}$	attribute
A	action partition of an extensive game
a	action
$a(x)$	action containing node x
A_h	action set at information set h
A_i	action set of player i
$\mathbf{a}_i$	pure strategy of i in an extensive game
$\mathbf{A}_i$	set of i's pure strategies in an extensive game
$\mathbf{a}$	pure strategy combination in an extensive game
$\mathbf{A}$	set of pure strategy combinations in an extensive game
$\mathbf{a}^+$	extended pure strategy combination in an extensive game
$\mathbf{A}^+$	set of extended pure strategy combinations in an extensive game
$\mathbf{a}_h^+$	action at h in $\mathbf{a}^+$
$\mathbf{A}^+(x)$	set of extended action combinations that lead to terminal nodes succeeding x
$\mathbf{A}^+(h)$	set of extended action combinations that lead to terminal nodes succeeding h
$\mathbf{AC}$	action condition for strong isomorphisms of extensive games
$a^j[s_i]$	action representing s_i in $\Gamma^j(G)$
b_h	local strategy at information set h
b_i	behavior strategy of player i
B_i	set of player i's behavior strategies
b	behavior strategy combination
B	set of behavior strategy combinations
B^0	set of completely mixed behavior strategy combinations
$c_i(S_i')$	centroid of player-index i with regard to S_i'
c_i	centroid of player-index i
$c_{I'}$	centroid of the players in I'
c	centroid of G
$\mathbf{CPL}$	action-chance mechanism condition for weak isomorphisms of extensive games

CPR	chance mechanism condition for isomorphisms of extensive games
CP	chance mechanism condition for strong isomorphisms of extensive games
FEG	framed extensive game
FSG	framed strategic game
f	isomorphism from G onto $\bar{G}$
f	bijective mapping $f : S \to \bar{S}$, induced by **f**
F	frame of a strategic game
$\mathcal{F}$	class of finite framed strategic games
$(\mathbf{f}, \mathbf{t})$	isomorphism of a FSG
F_r	frame of the FSG representing $\bar{G}^r$
f	strong isomorphism of extensive games (Chapter 3)
Φ	frame of an extensive game
$\mathcal{F}^e$	class of finite framed extensive games
G	strategic game
$\mathcal{G}$	class of finite strategic games
$\bar{G}^r$	M_n as perceived by a player with reptertoire r in a VUMG
red(G)	reduced form of G
red$^*(G)$	fully reduced form of G
(G, F)	framed strategic game
(G_r, F_r)	FSG representing $\bar{G}^r$
red(G, F)	reduced form of (G, F)
red$^*(G, F)$	fully reduced form of (G, F)
$\mathcal{G}^e$	class of finite extensive games with perfect recall
$G(\Gamma)$	agent normal form of Γ
Γ^x	subgame of Γ at node x
$\Gamma^j(G)$	j-orderd extensive representation of G
Γ	extensive game
Γ^{CH}	extensive game for Crawford & Haller's example
$(G(\Gamma), F)$	simply labelled agent normal form of Γ
$(G(\Gamma), F_{\text{hist}})$	history labelled agent normal form of Γ
(Γ, Φ)	framed extensive game
H	information partition of an extensive game
h	information set
$h(x)$	information set containing node x
hist(h)	history of information set h
HIS	history condition for isomorphisms of extensive games
i, i'	player-indices
I	set of player-indices
I', I''	subsets of I
$-i$	$I \backslash \{i\}$
$-I'$	$I \backslash I'$
$\bar{I}^r$	player set of $\bar{G}^r$

ISA	action-information set condition for weak isomorphisms of extensive games
IS	information set condition for strong isomorphisms of extensive games
(i, z_1, z_2)	information set in Γ^{CH}
(i, z_1, z_2, z)	action in Γ^{CH}
$\jmath$	order of players
L	solution concept
$L(G)$	solutions of the strategic game G
$\ell_{i\mathfrak{a}}$	label function of i concerning attribute $\mathfrak{a}$ in FSGs
λ	labels
Λ	label set of a frame
$L(G, F)$	solution of (G, F)
$L(\Gamma, \Phi)$	solution of (Γ, Φ)
$\ell_{\mathfrak{a}}$	label function concerning attribute $\mathfrak{a}$ in FEGs
M_n	Two-Player-One-Shot motching game with n options
μ	bijective mapping $\mu : \mathfrak{A} \to \overline{\mathfrak{A}}$, part of isomorphisms of FSGs
μ	system of beliefs in sequential equilibria
(μ, b)	assessment
$\mu(b)$	system of beliefs that is associated with b by Bayes' rule
$\mathbb{N}_+$	set of positive integers
$\mathbb{N}$	set of non-negative integers
$N(x)$	set of successors of x
o	root of an extensive game
π	part of an isomorphism, bijective mapping $\pi : I \to \bar{I}$
P	player partition of an extensive game
p	chance mechanism of an extensive game
$\psi(x)$	path of x
$\dot{\psi}(x)$	ordered path of x
P_i	cell of player i in P
$p_h(a)$	probability of action a at information set h of the chance mechanism
$\text{prob}(\mathbf{a}^+|\sigma)$	conditional probability of $\mathbf{a}^+$ given σ
$\text{prob}(z'|\sigma)$	conditional probability of z' given σ
$\text{prob}(\mathbf{a}^+|b)$	conditional probability of $\mathbf{a}^+$ given b
$\text{prob}(z'|b)$	conditional probability of z' given b
$\text{prob}(z|b, x)$	probability to reach z from x by b
PL	condition concerning the action-player structure for weak
P	player condition for strong isomorphisms of extensive games
PTH	path condition for weak isomorphisms of extensive games
PTH$^+$	strong path condition for isomorphisms of extensive games
PTH$^-$	weak path condition for isomorphisms of extensive games

PY	payoff condition for weak isomorphisms of extensive games
PY*	payoff condition for strong isomorphisms of extensive games
q_1, q_2	permutations defining weak symmetries of extensive games
$\mathbb{R}$	real line
r_i	part of an isomorphism, bijective mapping $r_i : S_i \to \bar{S}_{\pi(i)}$
R	set of repertoires
r	repertoire (Chapter 2)
r	weak isomorphism of extensive games (Chapter 3)
$\mathbf{r}^+$	bijection $\mathbf{r}^+ : \mathbf{A}^+ \to \bar{\mathbf{A}}^+$, induced by r
$\mathbf{r}$	bijection $\mathbf{r} : \mathbf{A} \to \bar{\mathbf{A}}$, induced by r
$(r, \mathbf{t})$	weak isomorphism of FEGs
s_i	pure strategy-index of player-index i
s_{ik}	kth pure strategy of i
S_i	set of i's pure strategy-indices
S	set of pure strategy-index combinations
σ_i	mixed strategy-index of player-index i
Σ_i	set of i's mixed strategy-indices
$\sigma_i(s_i)$	probability of i's pure strategy-index s_i in σ_i
σ	mixed strategy-index combination
Σ	set of mixed strategy-index combinations
$S_{I'}$	set of pure I'-subcombinations
$s_{I'}$	pure I'-subcombination
$\Sigma_{I'}$	set of mixed I'-subcombinations
$\sigma_{I'}$	mixed I'-subcombination
S_i'	subset of S_i
Σ_i^0	set of i's completely mixed strategy-indices
Σ^0	set of completely mixed strategy-index combinations
$\mathfrak{S}_i$	set of equivalence classes of S_i induced by $\sim_i$
$\mathfrak{s}_i(s_i)$	class of $\mathfrak{S}_i$ containing s_i
$S_i(r)$	perceived-option set of player i with repertoire r
$S_i(\mathfrak{a}, \lambda)$	set of i's pure strategy-indices that are assigned the label λ with regard to the attribute $\mathfrak{a}$
$[s_i]_r$	equivalence class of s_i concerning $\sim_{i,r}$
$\bar{S}_i^r$	pure strategy set of i in $\bar{G}^r$
$\Sigma_i(r)$	set of mixed strategy-indices of i with repertoire r in $\bar{G}^r$
$\Sigma(r)$	set of mixed strategy-index combinations in $\bar{G}^r$
$\bar{\theta}$	strategy-index combination in a VUMG
$\bar{\Theta}$	set of strategy-index combinations in a VUMG
$\mathbf{t}$	$= (\mu, \tau)$, part of an isomorphism of FSGs
θ	strategy-index combination in the FSG representation of a VUMG
Θ	set of strategy-index combinations in the FSG representation of a VUMG

τ	bijective mapping $\tau : \Lambda \to \bar{\Lambda}$, part of isomorphisms of FSGs
T	node set of an extensive game
θ	bijective mapping $\theta : Z \to \bar{Z}$, part of weak isomorphisms of extensive games
TR	tree condition for strong isomorphisms of extensive games
u_ι	payoff function
$u_i(\sigma)$	player-index' i's payoff of strategy-index combination σ
$\bar{u}_i^r$	payoff function of i with repertoire r in a VUMG
u_i^r	payoff function of i with repertoire r in the FSG representation of a VUMG
u	payoff function of an extensive game
$u_i(z)$	payoff of player i at the terminal node z
$u_h(\mu, b)$	payoff of the player at h with regard to the belief system μ
$u_x(b)$	payoff of the player at x
VFT	variable frame theory of focal points
$v(r)$	probability of repertoire r in v
v	distribution of repertoires
$v(r)$	probability of repertoire r in v
$v(r'|r)$	expectation of a player with repertoire r that the other player has repertoire r'
VUMG	variable universe matching game
$V(x)$	predecessor of node x
ν	bijection $\nu : H \to \bar{H}$, implied by weak isomorphisms of extensive games
x, x', x''	nodes of an extensive game
X	set of decision nodes of an extensive game
Z	set of terminal nodes of an extensive game
$z(\mathbf{a}^+)$	terminal node reached by $\mathbf{a}^+$
$Z(x)$	terminal nodes succeeding x
$(z_{11}, z_{12}, z_{21}, z_{22})$	terminal node in Γ^{CH}

List of Figures

Bibliography

Agassi, A. (1960). Methodological individualism, *British Journal of Sociology* **11**: 144–170.

Anderlini, L. (1999). Communication, computability, and common interest games, *Games and Economic Behavior* **27**: 1–37.

Arrow, K. J. & Hurwicz, L. (1972). An optimality criterion for decision-making under ignorance, *in* C. F. Carter & J. L. Ford (eds), *Uncertainty and Expectations in Economics*, Basil Blackwell, Oxford, pp. 1–11.

Aumann, R. J. (1976). Agreeing to disagree, *Annals of Statistics* **4**: 1236–1239.

Bacharach, M. (1991). Games with concept-sensitive strategy spaces, Oxford University, mimeo.

Bacharach, M. (1993). Variable universe games, *in* K. Binmore, A. Kirman & P. Tani (eds), *Frontiers of Game Theory*, MIT Press, Cambridge, Mass., pp. 255–275.

Bacharach, M. (1998). Players' representations and the theory of games with multiple identity, Working paper, Institute of Economics and Statistics, University of Oxford.

Bacharach, M. & Bernasconi, M. (1997). The variable frame theory of focal points: An experimental study, *Games and Economic Behavior* **19**: 1–45.

Bacharach, M. & Stahl, D. O. (2000). Variable-frame level-n theory, *Games and Economic Behavior* **32**: 220–246.

Bernheim, D. (1984) Rationalizable strategic behavior, *Econometrica* **52**(4): 1007–1028.

Binmore, K., Swierzbinski, J., Hsu, S. & Proulx, C. (1993). Focal points and bargaining, *International Journal of Game Theory* **22**: 381–409.

Blume, A. & Gneezy, U. (1998). An experimental investigation of optimal learning in games, Working paper, Departments of Economics, University of Iowa and University of Haifa.

Border, K. C. (1989). *Fixed Point Theorems with Applications to Economics and Game Theory*, Cambridge University Press, Cambridge.

Casajus, A. (1997) Coordination, salience, and symmetries in framed standard forms, Paper presented at the 17th Arne Ryde Symposium, August 21-22, 1997, Lund, Sweden

Casajus, A (1998). Focal points in framed extensive forms, Paper presented at the 3rd Spanish Meeting on Game Theory and Applications, June 15-17, 1998, Barcelona, Spain.

Crawford, V. P. & Haller, H. (1990). Learning how to cooperate: Optimal play in repeated coordination games, *Econometrica* **58**(3): 571–595.

De Francesco, M. A. (1996). Customer markets as an efficient outcome of a dynamic choosing-the-seller game, Quaderni del Dipartimento di Econimica Politica 205, Universita Degli Studi de Siena, Italy.

Elmes, S. & Reny, P. J. (1994). On the strategic equivalence of extensive form games, *Journal of Economic Theory* **62**(1): 1–23.

Gauthier, D. (1975). Coordination, *Dialogue* **14**: 195–221.

Gilbert, M. (1989). Rationality and salience, *Philosophical Studies* **57**: 61–77.

Gilbert, M. (1990). Rationality, coordination, and convention, *Synthese* **84**: 1–21.

Goyal, S. & Janssen, M. C. W. (1996). Can we rationally learn to coordinate?, *Theory and Decision* **40**(1): 29–49.

Grant, S. & Quiggin, J. (1998). The meeting place problem: Salience and search, *Journal of Economic Behavoir and Organization* **33**: 271–283.

Harsanyi, J. C. & Selten, R. (1988). *A General Theory of Equilibrium Selection in Games*, MIT Press, Cambridge, Mass., London, England.

Hildenbrand, W. & Kirman, A. P. (1988). *Equlibrium Analysis*, North-Holland, Amsterdam u.a.

Hollis, M. & Sugden, R. (1993). Rationality in action, *Mind* **102**(405): 1–35.

Janssen, M. C. W. (1997). Rationalizing focal points, Discussion Paper, Erasmus University Rotterdam.

Kahneman, D. & Tversky, A. (1984). Choices, values, and frames, *American Psychologist* **39**(4): 341–350.

Kalai, E. & Samet, D. (1984). Persistent equilibria, *International Journal of Game Theory* **13**: 129–144.

Kandori, M., Mailath, G. J. & Rob, R. (1993). Learning, mutation, and long run equilibria in games, *Econometrica* **61**(1): 29–56.

Kohlberg, E. & Mertens, J.-F. (1986). On the strategic stability of equilibria, *Econometrica* **54**(5): 1003–1037.

Kramarz, F. (1996). Dynamic focal-points in n-person coordination games, *Theory and Decision* **40**(3): 277–313.

Kreps, D. M. (1988). *Notes on the Theory of Choice*, Westview Press, Boulder and London.

Kreps, D. M. & Wilson, R. (1982). Sequential equilibria, *Econometrica* **50**(4): 863–94.

Kuhn, H. W. (1953). Extensive games and the problem of information, *in* H. W. Kuhn & A. W. Tucker (eds), *Contributions to the Theory of Games*, Princeton University Press, Princeton, pp. 193–216.

Lewis, D. K. (1969). *Convention*, Harvard University Press, Cambridge, Mass.

Mehta, J., Starmer, C. & Sugden, R. (1991). An experimental investigation of focal points in coordination and bargaining: Some preliminary results, Dicussion Paper No. 9101, School of Economic and Social Studies, University of East Anglia.

Mehta, J., Starmer, C. & Sugden, R. (1994a). Focal points in pure coordination games: An experimental investigation, *Theory and Decision* **36**: 163–185.

Mehta, J., Starmer, C. & Sugden, R. (1994b). The nature of salience: An experimental investigation of pure coordination games, *American Economic Review* **84**(3): 658–673.

Miller, S. (1991). Co-ordination, salience and rationality, *Southern Journal of Philosophy* **XXIX**(3): 359–370.

Milnor, J. (1954). Games against nature, *in* R. M. Thrall, C. H. Coombs & R. L. Davis (eds), *Decision Processes*, Wiley, New York, pp. 49–59.

Morgenbesser, S. & Ullmann-Margalit, E. (1977). Picking and choosing, *Social Research* **44**(4): 757–785.

Myerson, R. B. (1978). Refinements of the Nash equilibrium concept, *International Journal of Game Theory* **7**(2): 73–80.

Myerson, R. B. (1991). *Game Theory: Analysis of Conflict*, Harvard University Press, Cambridge, Mass., London.

Nash, J. F. (1950). Equilibrium points in n-person games, *Proceedings of the National Academy of Science U.S.A.* **36**: 48–49.

Nash, J. F. (1951). Non-cooperative games, *Annals of Mathematics* **54**(2): 286–295.

Oh, J. H. (1995). *Three Essays on Equilibrium Selection in Games*, PhD thesis, Graduate College of The University of Iowa.

Osborne, M. J. & Rubinstein, A. (1994). *A Course in Game Theory*, MIT Press, Cambridge, MA, London, England.

Pearce, D. G. (1984). Rationalizable strategic behavior and the problem of perfection, *Econometrica* **52**(4): 1029–1050.

Peleg, B., Rosenmüller, J. & Sudhölter, P. (1999). The canonical extensive form of a game form: Symmetries, *in* A. Alkan, C. Aliprantis & N. Yannelis (eds), *Current Trends in Economics: Theory and Applications*, Springer, pp. 367–387.

Peleg, B., Rosenmüller, J. & Sudhölter, P. (2000). The canonical extensive form of a game form: Representation, *Journal of Mathematical Economics* **33**(3): 299–338.

Piccione, M. & Rubinstein, A. (1997). On the interpretation of decision problems with imperfect recall, *Games and Economic Behaviour* **20**: 3–24.

Provis, C. (1977). Gauthier on coordination, *Dialogue* **16**(3): 507–509.

Roth, A. E. (1985). Toward a focal point theory of bargaining, *in* A. E. Roth (ed.), *Game-Theoretic Models of Bargaining*, Cambridge University Press, Cambridge, pp. 259–268.

Roth, A. E. (1995). Bargaining experiments, *in* J. H. Kagel & A. E. Roth (eds), *The Handbook of Experimental Economics*, Princeton University Press, pp. 253–348.

Rottenstreich, Y. S. (1996). *Framing in Games*, PhD thesis, Stanford University.

Rubinstein, A. & Tversky, A. (1993). Naive strategies in zero-sum games, Working Paper No. 17-93, The Sackler Instiutute of Economic Studies, Tel-Aviv University, Israel.

Schelling, T. C. (1960). *The Strategy of Conflict*, 8 edn, Harvard University Press, Cambridge, Mass., London, England.

Schlag, K. H. (1993). Cheap talk and evolutionary dynamics, Discussion Paper No. B-242, University of Bonn, Germany.

Selten, R. (1975). Reexamination of the perfectness concept for equilibrium points in extensive games, *International Journal of Game Theory* **4**(1): 25–55.

Selten, R. (1983). Evolutionary stability in extensive two-person games, *Mathematical Social Sciences* **5**: 269–363.

Selten, R. (1988). Evolutionary stability in extensive two-person games - correction and further development, *Mathematical Social Sciences* **16**: 223–266.

Shubik, M. (1994). Prominence, symmetry, or other?, *Games and Economic Behavior* **6**: 328–332.

Sugden, R. (1991). Rational choice: A survey of contributions from economics and philosophy, *Economic Journal* **101**: 751–785.

Sugden, R. (1993). Thinking as a team: Towards an explanation of nonselfish behavior, *Social Philosophy & Policy* **10**(1): 69–89.

Sugden, R. (1995). A theory of focal points, *Economic Journal* **105**: 533–550.

Thompson, F. B. (1952). Equivalence of games in extensive form, RM 759, The Rand Corporation.

Tversky, A. & Kahneman, D. (1981). The framing of decisions and the psychology of choice, *Science* **211**: 453–458.

Young, H. P. (1993). The evolution of conventions, *Econometrica* **61**(1): 57–84.

Young, H. P. (1996). The economics of convention, *Journal of Economic Perspectives* **10**(2): 105–122

Lecture Notes in Economics and Mathematical Systems

For information about Vols. 1–305
please contact your bookseller or Springer-Verlag